Mathematical Algorithms in Visual Basic for Scientists & Engineers

Other McGraw-Hill Titles of Interest

Mathematical Algorithms in Visual Basic for Scientists & Engineers

Namir C. Shammas

McGraw-Hill, Inc.

New York San Francisco Washington, D.C. Auckland Bogotá
Caracas Lisbon London Madrid Mexico City Milan
Montreal New Delhi San Juan Singapore
Sydney Tokyo Toronto

McGraw-Hill

A Division of The McGraw·Hill Companies

Library of Congress Cataloging-in-Publication Data

Shammas, Namir Clement, 1954–
    Mathematical algorithms in Visual Basic for scientists & engineers
/ by Namir C. Shammas.
        p. cm.
    Includes index.
    ISBN 0-07-912003-2 (pbk.)
    1. Numerical analysis—Data processing. 2. Mathematical
statistics—Data processing. 3. Microsoft Visual Basic for Windows.
I. Title.
QA297.S4585 1995
519.4'0285'513—dc20   95-32896
                                                          CIP

pbk 1 2 3 4 5 6 7 8 9 FGR/FGR 9 0 0 8 7 6 5

ISBN 0-07-912003-2

The sponsoring editor was Jennifer Holt DiGiovanna, the executive editor was Joanne Slike, the associate managing editor was David M. Mc-Candless, and the book editor of this book was Marianne Krcma and the production supervisor was Katherine G. Brown. This book was set in ITC Century Light. It was composed in Blue Ridge Summit, Pennsylvania.

Printed and bound by Quebecor, Fairfield, PA.

McGraw-Hill books are available at special quantity discounts to use as premiums and sales promotions, or for use in corporate training programs. For more information, please write to the Director of Special Sales, McGraw-Hill, 11 West 19th Street, New York, NY 10011. Or contact your local bookstore.

MH95
9120032

To a modern-day Helen of Troy, Anna Bazaco

Contents

Introduction

For many decades, Fortran has monopolized statistical and numerical analysis computing. As Visual Basic gains popularity, many find this Windows-based BASIC dialect fairly suitable for the same tasks.

This book covers popular numerical analysis and statistical topics implemented in Visual Basic. Wherever relevant, the algorithms involved in the methods presented are also included. As for the equations behind the methods, they are of secondary importance. Many books discuss the equations behind the various numerical and statistical methods. However, not many books present pseudocode for the algorithms. This book does, allowing you to use the pseudocode to develop libraries in other programming languages, such as C and Pascal.

The twelve chapters in this book are arranged as follows:

- Chapter 1 discusses simultaneous linear equations. The chapter presents a library for general vectors and matrices, and then discusses popular methods for solving linear equations.

- Chapter 2 looks at solving nonlinear equations. The text discusses various methods for solving the roots of functions with single and multiple variables.

- Chapter 3 focuses on interpolation methods, including Lagrangian interpolation, the Barycentric interpolation, Newton's divided difference interpolation, the Newton difference method, and the cubic spline method.

- Chapter 4 discusses numerical differentiation and presents the forward/backward difference method, the central difference method, and the extended central difference method. These methods offer you nontrivial tools for estimating the derivative of a function.

- Chapter 5 looks at popular numerical integration methods such as Simpson's method, the Gaussian quadrature methods, and the Romberg method. These methods offer tools for calculating the finite and infinite integrals for functions.

- Chapter 6 focuses on solving ordinary differential equations. The chapter discusses the popular Runge-Kutta method and two of its spinoff methods, Runge-

Kutta-Gill and Runge-Kutta-Fehlberg. These spinoff methods offer improved accuracy for the solutions.

- Chapter 7 discusses the optimization of single-variable and multivariable functions. The chapter discusses methods such as the Golden Section search method, interpolation methods, and the simplex method.

- Chapter 8 looks at basic statistics, including the mean, standard deviation, the confidence intervals for the mean and standard deviation, and the first four moments. The chapter also looks at testing sample means to determine whether or not they are significantly different.

- Chapter 9 discusses the various ANOVA tests. These tests include the one-way ANOVA, the two-way ANOVA, the two-way ANOVA with replication, the Latin-square ANOVA, and the analysis of covariance.

- Chapter 10 presents a library that supports linear regression. This chapter covers basic linear regression, linearized regression, the confidence interval for projections, and the confidence interval for regression coefficients. The text also discusses automatic best fit for linearized regression.

- Chapter 11 discusses multiple and polynomial regression. The text looks at calculating the regression intercept, slopes, standard errors for the slopes, projections, the confidence intervals for the slopes, and performing various Student-t tests for the slopes and correlation coefficient.

- Chapter 12 presents libraries for common statistical and mathematical functions. The statistical functions library includes the normal distribution, Student-t distribution, the F distribution, and their inverses. The mathematical functions library includes functions such as the combination function, the gamma function, the error function, and the Laguerre functions.

I hope that this book offers you a time-saving library for common methods in numerical analysis and statistics.

Happy Programming!

Simultaneous Linear Equations

Solving simultaneous linear equations is among the most popular operations in numerical analysis, as well as other scientific, statistical, and engineering fields. This chapter presents selected methods for solving linear equations in Visual Basic. The chapter presents and discusses the following topics:

- The Gauss-Jordan elimination method
- The Gauss-Seidel method
- The LU decomposition method
- The Visual Basic source code for the above methods
- The Visual Basic test program

The Visual Basic listings in this chapter solve simultaneous linear equations for the above methods and perform miscellaneous matrix operations.

☞ Throughout the book, the underscore character is used to split wrapping lines of Visual Basic declarations and statements.

The Gauss-Jordan Elimination Method

The subject of solving simultaneous linear equations has occupied mathematicians for a long time. Today, we have a repertoire of numerous methods that vary in approach and sophistication. This section discuss one of the popular methods for solving linear equations, namely, the Gauss-Jordan elimination method. This method is generally as efficient as other methods and is stable. The algorithm implemented in this chapter uses full pivoting in the process of reducing the set of linear equations.

The Gauss-Seidel Method

The Gauss-Seidel method is an iterative technique that solves a set of linear equations by providing an initial guess and then refining that guess. Each iteration cycle calculates the updated guess for each variable in terms of the current values of the other variable. The algorithm employs the updated values for a variable within an iteration, instead of waiting for the current iteration to complete.

The LU Decomposition Method

The LU decomposition method solves linear equations by performing two main steps. The first step transforms the main matrix into upper/lower triangular matrices. The second step solves for a specific solution vector using backsubstitution. You can repeat the second steps as many times as needed to solve for different solution vectors.

The Visual Basic Source Code

Let's look at the Visual Basic source code that implements the above linear equations algorithms, as well as other algorithms that manipulate matrices. Listing 1.1 shows the source code for the MATVECT.BAS module file.

Listing 1.1 The source code for the MATVECT.BAS module file.

```
Global Const MATVECT_EPSILON# = .000000000000001
Global Const MATVECT_BAD_RESULT# = -1E+30
Global Const matErr_None% = 0
Global Const matErr_Size% = 1
Global Const matErr_Singular% = 2
Global Const matErr_IllConditioned% = 3
Global Const matErr_IterLimit% = 4

Function AddMat (MatC() As Double, MatA() As Double, _
 MatB() As Double, numRows As Integer, numCols As Integer) _
 As Integer

 Dim row As Integer, col As Integer

 If Not (checkRowCol(MatA(), numRows, numCols) And _
     checkRowCol(MatB(), numRows, numCols) And _
     checkRowCol(MatC(), numRows, numCols) And _
     (numCols = numRows)) Then
  AddMat = matErr_Size
  Exit Function
 End If

 For row = 0 To numRows - 1
  For col = 0 To numCols - 1
   MatC(row, col) = MatA(row, col) + MatB(row, col)
  Next col
 Next row
 AddMat = matErr_None
End Function

Function checkRowCol (Mat() As Double, row As Integer, _
 col As Integer) As Integer
 If (row >= 0) And (col >= 0) And _
   (row < UBound(Mat, 1)) And _
   (col < UBound(Mat, 2)) Then
```

Listing 1.1 (*Continued*)

```
  checkRowCol = True
 Else
  checkRowCol = False
 End If
End Function

Function CopyMat (MatB() As Double, MatA() As Double, _
 numRows As Integer, numCols As Integer) As Integer

 Dim row As Integer, col As Integer

 If Not (checkRowCol(MatA(), numRows, numCols) And _
     checkRowCol(MatB(), numRows, numCols)) Then
   CopyMat = matErr_Size
   Exit Function
 End If

 For row = 0 To numRows - 1
  For col = 0 To numCols
   MatB(row, col) = MatA(row, col)
  Next col
 Next row
 CopyMat = matErr_None
End Function

Function GaussJordan (A() As Double, B() As Double, _
 numRows As Integer, numCols As Integer) As Integer

 Dim rowIndex() As Integer
 Dim colIndex() As Integer
 Dim pivotIndex() As Integer
 Dim i As Integer, j As Integer, k As Integer
 Dim n As Integer, m As Integer
 Dim row As Integer, col As Integer
 Dim large As Double, z As Double, oneOverPiv As Double

 If Not checkRowCol(A(), numRows, numCols) Then
  GaussJordan = matErr_Size
  Exit Function
 End If

 ' resize local dynamic arrays
 ReDim rowIndex(numRows)
 ReDim colIndex(numRows)
 ReDim pivotIndex(numRows)

 ' initialize the row and column indices
 For i = 0 To numRows - 1
  rowIndex(i) = i
  colIndex(i) = i
 Next i

 ' initialize the pivot index array
 For i = 0 To numRows - 1
  pivotIndex(i) = -1
 Next i

 For i = 0 To numRows - 1
  large = 0
  For j = 0 To numRows - 1
   If pivotIndex(j) <> 0 Then
    For k = 0 To numRows - 1
     If pivotIndex(k) = -1 Then
```

Listing 1.1 (*Continued*)

```
      If Abs(A(j, k)) >= large Then
       large = Abs(A(j, k))
       row = j
       col = k
      End If
     ElseIf pivotIndex(k) > 0 Then
      GaussJordan = matErr_Singular
      Exit Function
     End If
    Next k
    End If
   Next j
   pivotIndex(col) = pivotIndex(col) + 1
   If row <> col Then
    For n = 0 To numRows - 1
     swapDouble A(row, n), A(col, n)
    Next n
    For n = 0 To numCols - 1
     swapDouble B(row, n), B(col, n)
    Next n
   End If
   rowIndex(i) = row
   colIndex(i) = col
   If Abs(A(col, col)) < .0000000001 Then
    GaussJordan = matErr_Singular
    Exit Function
   End If
   oneOverPiv = 1 / A(col, col)
   A(col, col) = 1
   For n = 0 To numRows - 1
    A(col, n) = A(col, n) * oneOverPiv
   Next n
   For n = 0 To numCols - 1
    B(col, n) = B(col, n) * oneOverPiv
   Next n
   For m = 0 To numRows - 1
    If m <> col Then
     z = A(m, col)
     A(m, col) = 1
     For n = 0 To numRows - 1
      A(m, n) = A(m, n) - A(col, n) * z
     Next n
     For n = 0 To numCols - 1
      B(m, n) = B(m, n) - B(col, n) * z
     Next n
    End If
   Next m
  Next i
  For n = numRows - 1 To 0 Step -1
   If rowIndex(n) <> colIndex(n) Then
    For k = 0 To numRows - 1
     swapDouble A(k, rowIndex(n)), A(k, colIndex(n))
    Next k
   End If
  Next n
  GaussJordan = matErr_None
End Function

Function GaussSeidel (A() As Double, B() As Double, _
 X() As Double, numRows As Integer, maxIter As Integer, _
 eps1 As Double, eps2 As Double)
```

Listing 1.1 (*Continued*)

```
Const opContinue% = 1
Const opConverge% = 2
Const opSingular% = 3
Const opError% = 4

Dim Xold() As Double
Dim denom As Double, sum As Double
Dim dev As Double, devMax As Double
Dim i As Integer, j As Integer
Dim iter As Integer
Dim operType As Integer

iter = 0
operType = opContinue
ReDim Xold(numRows)
' normalize matrix A and vector B
For i = 0 To numRows - 1
 denom = A(i, i)
 If denom < eps1 Then
  GaussSeidel = matErr_Singular
  Exit Function
 End If
 B(i) = B(i) / denom
 For j = 0 To numRows - 1
  A(i, j) = A(i, j) / denom
 Next j
Next i

' perform Gauss-Seidel iteration
Do While operType = opContinue
 For i = 0 To numRows - 1
  Xold(i) = X(i)
  X(i) = 0
  For j = 0 To numRows - 1
  If j <> i Then
   X(i) = X(i) - A(i, j) * X(j)
  End If
  Next j
  X(i) = X(i) + B(i)
 Next i

 ' check for the convergence
 devMax = Abs(Xold(0) - X(0)) / X(0)
 For i = 1 To numRows - 1
  dev = Abs(Xold(i) - X(i)) / X(i)
  If dev > devMax Then devMax = dev
 Next i
 If devMax <= eps2 Then
  operType = opConverge
 Else
  iter = iter + 1
  If iter > maxIter Then operType = opError
 End If
Loop

Select Case operType
 Case Is = opConverge
  GaussSiedel = matErr_None

 Case Is = opSingular
  GaussSiedel = matErr_Singular
```

Listing 1.1 (*Continued*)

```
  Case Is = opError
   GaussSiedel = matErr_IterLimit

  Case Else
   GaussSiedel = matErr_None
 End Select
End Function

Sub LUBackSubst (A() As Double, Index() As Integer, _
 numRows As Integer, B() As Double)

 Dim i As Integer, j As Integer
 Dim idx As Integer, k As Integer
 Dim sum As Double

 k = -1
 For i = 0 To numRows - 1
  idx = Index(i)
  sum = B(idx)
  B(idx) = B(i)
  If k > -1 Then
   For j = k To i - 1
    sum = sum - A(i, j) * B(j)
   Next j
  ElseIf sum <> 0 Then
   k = i
  End If
  B(i) = sum
 Next i
 For i = numRows - 1 To 0 Step -1
  sum = B(i)
  For j = i + 1 To numRows - 1
   sum = sum - A(i, j) * B(j)
  Next j
  B(i) = sum / A(i, i)
 Next i
End Sub

Function LUDecomp (A() As Double, Index() As Integer, _
 numRows As Integer, rowSwapFlag As Integer) As Integer

 Dim i As Integer, j As Integer
 Dim k As Integer, iMax As Integer
 Dim large As Double, sum As Double
 Dim z As Double, z2 As Double
 Dim scaleVect() As Double

 ReDim scaleVect(UBound(A, 2))
 ' initialize row interchange flag
 rowSwapFlag = 1
 ' loop to obtain the scaling element
 For i = 0 To numRows - 1
  large = 0
  For j = 0 To numRows - 1
   z2 = Abs(A(i, j))
   If z2 > large Then large = z2
  Next j
  ' no non-zero large value? then exit with an error code
  If large = 0 Then
   LUDecomp = matErr_Singular
   Exit Function
  End If
```

Listing 1.1 (*Continued*)

```
  scaleVect(i) = 1 / large
 Next i
 For j = 0 To numRows - 1
  For i = 0 To j - 1
   sum = A(i, j)
   For k = 0 To i - 1
   sum = sum - A(i, k) * A(k, j)
   Next k
   A(i, j) = sum
  Next i
  large = 0
  For i = j To numRows - 1
   sum = A(i, j)
   For k = 0 To j - 1
   sum = sum - A(i, k) * A(k, j)
   Next k
   A(i, j) = sum
   z = scaleVect(i) * Abs(sum)
   If z >= large Then
   large = z
   iMax = i
   End If
  Next i
  If j <> iMax Then
   For k = 0 To numRows - 1
   z = A(iMax, k)
   A(iMax, k) = A(j, k)
   A(j, k) = z
   Next k
   rowSwapFlag = -1 * rowSwapFlag
   scaleVect(iMax) = scaleVect(j)
  End If
  Index(j) = iMax
  If A(j, j) = 0 Then
   A(j, j) = MATVECT_EPSILON
  End If
  If j <> numRows Then
   z = 1 / A(j, j)
   For i = j + 1 To numRows - 1
   A(i, j) = A(i, j) * z
   Next i
  End If
 Next j
 LUDecomp = matErr_None
End Function

Function LUDeterminant (A() As Double, numRows As Integer, _
 rowSwapFlag As Integer) As Double

 Dim result As Double
 Dim i As Integer

 result = rowSwapFlag
 For i = 0 To numRows - 1
  result = result * A(i, i)
 Next i

 LUDeterminant = result
End Function

Sub LUInverse (A() As Double, InvA() As Double, _
 Index() As Integer, numRows As Integer)
```

Listing 1.1 *(Continued)*

```
Dim colVect() As Double
Dim i As Integer, j As Integer

ReDim colVect(numRows)
For j = 0 To numRows - 1
 For i = 0 To numRows - 1
  colVect(i) = 0
 Next i
 colVect(j) = 1
 LUBackSubst A(), Index(), numRows, colVect()
 For i = 0 To numRows - 1
  InvA(i, j) = colVect(i)
 Next i
Next j
End Sub

Function MatDeterminant (A() As Double, numRows As Integer) _
 As Double

 Dim Index() As Integer
 Dim i As Integer, j As Integer
 Dim rowSwapFlag As Integer
 Dim fxErr As Integer
 Dim result As Double

 ReDim Index(numRows)
 fxErr = LUDecomp(A(), Index(), numRows, rowSwapFlag)
 If Err <> matErr_None Then
  MatDeterminant = MATVECT_BAD_RESULT
  Exit Function
 End If

 result = rowSwapFlag

 For i = 0 To numRows - 1
  result = result * A(i, i)
 Next i
 MatDeterminant = result
End Function

Function MatInverse (A() As Double, numRows As Integer) _
 As Integer

 Dim colVect() As Double
 Dim Index() As Integer
 Dim i As Integer, j As Integer
 Dim rowSwapFlag As Integer
 Dim fxErr As Integer

 ReDim colVect(numRows)
 ReDim Index(numRows)
 Err = LUDecomp(A(), Index(), numRows, rowSwapFlag)
 If Err <> matErr_None Then
  MatInverse = Err
  Exit Function
 End If

 For j = 0 To numRows - 1
  For i = 0 To numRows - 1
   colVect(i) = 0
  Next i
  colVect(j) = 1
  LUBackSubst A(), Index(), numRows, colVect()
```

Listing 1.1 (*Continued*)

```
  For i = 0 To numRows - 1
   A(i, j) = colVect(i)
  Next i
 Next j

 MatInverse = matErr_None
End Function

Function MulMat (MatC() As Double, MatA() As Double, _
 MatB() As Double, numRows As Integer, numCols As Integer) _
 As Integer

 Dim row As Integer, col As Integer, k As Integer

 If Not checkRowCol(MatA(), numRows, numCols) Then
  MulMat = matErr_Size
  Exit Function
 End If

 If Not checkRowCol(MatB(), numRows, numCols) Then
  MulMat = matErr_Size
  Exit Function
 End If

 If numColsA <> numRowsB Then
  MulMat = matErr_Size
  Exit Function
 End If

 For row = 0 To numRowsB - 1
  For col = 0 To numColsA - 1
   MatC(row, col) = 0
   For k = 0 To numColsA - 1
   MatC(row, col) = MatC(row, col) + MatA(row, k) * _
           MatB(row, k)
   Next k
  Next col
 Next row
 MulMat = matErr_None
End Function

Function SubMat (MatC() As Double, MatA() As Double, _
 MatB() As Double, numRows As Integer, numCols As Integer) _
 As Integer

 Dim row As Integer, col As Integer

 If Not (checkRowCol(MatA(), numRows, numCols) And _
     checkRowCol(MatB(), numRows, numCols) And _
     checkRowCol(MatC(), numRows, numCols) And _
     (numCols = numRows)) Then
  SubMat = matErr_Size
  Exit Function
 End If

 For row = 0 To numRows - 1
  For col = 0 To numCols - 1
   MatC(row, col) = MatA(row, col) - MatB(row, col)
  Next col
 Next row
 SubMat = matErr_None
End Function
```

Listing 1.1 (*Continued*)

```
Sub swapDouble (d1 As Double, d2 As Double)

 Dim t As Double

 t = d1
 d1 = d2
 d2 = t
End Sub

Sub swapInt (i1 As Integer, i2 As Integer)
' swap two integers
 Dim t As Integer
 t = i1
 i1 = i2
 i2 = t
End Sub
```

Listing 1.1 declares a set of constants, most of which emulated enumerated values for the error code. The listing also declares a collection of functions and subroutines. The matErr_XXXX constants are associated with errors in solving linear equations. Many of the functions return values of the matErr_XXXX constants.

The listing contains a set of Visual Basic functions and subroutines that perform basic matrix manipulation, including these:

1. The function CopyMat copies matrix MatA into matrix MatB. The parameters numRows and numCols specify the number of rows and columns to copy.

2. The function AddMat adds matrices MatA and MatB, and stores the result in matrix MatC. The parameters numRows and numCols specify the number of rows and columns to add.

3. The function SubMat subtracts matrices MatA and MatB, and stores the result in matrix MatC. The parameters numRows and numCols specify the number of rows and columns to subtract.

4. The function MultMat multiplies matrices MatA and MatB, and stores the result in matrix MatC. The parameters numRows and numCols specify the number of rows and columns to multiply.

The module file MATVECT.BAS contains the following routines that solve for linear equations:

1. The function GaussJordan implements the Gauss-Jordan method. The parameter A is the matrix of coefficients. The parameter B is the constants matrix. The parameter numRows specifies the number of rows and columns in matrix A, and the number of rows in matrix B. The parameter numCols specifies the number of columns in matrix B and represents the number of solution sets required.

2. The function LUDecomp supports the LU decomposition of a matrix of coefficients, represented by parameter A. The parameter numRows specifies the number of rows and columns. The parameter Index is an integer vector that obtains the row indices. The parameter rowSwapFlag is the row-swap flag, which obtains its value from the function.

3. The subroutine LUBackSubst complements function LUDecomp by solving for a solution vector. The parameter A presents the LU matrix. The parameter Index is the vector of row indices. The parameter numRows specifies the number of rows and columns. The parameter B represents the constant vector (as input) and solution vector (as output).

4. The function GaussSeidel implements the Gauss-Seidel method. The parameter A represents the matrix of coefficients. The parameter B is the constants vector. The parameter X is the solution vector. The parameter numRows specifies the number of rows and columns. The parameters eps1 is the small value that is compared with the diagonal elements to determine if the matrix A is singular. The parameter eps2 is the small value used in testing for convergence.

5. The subroutine LUInverse yields the inverse matrix based on an LU matrix produced by function LUDecomp. The parameter A represents the LU matrix of coefficients. The parameter InVA represents the inverse matrix of A. The parameter Index is the index of rows. The parameter numRows specifies the number of rows (which is also equal to the number of columns).

6. The function MatInverse yields the inverse of a matrix. The parameter A represents the input matrix and the output inverse matrix. The parameter numRows specifies the number of rows (which is also equal to the number of columns).

7. The function LUDeterminant returns the determinant of a matrix. The parameter A is the matrix generated by the function LUDecomp. The parameter numRows specifies the number of rows (which is also equal to the number of columns). The parameter rowSwapFlag is the row-swap flag supplied by function LUDecomp.

8. The function MatDeterminant returns the determinant of a matrix. The parameter A is the input matrix. The parameter numRows specifies the number of rows (which is also equal to the number of columns).

The functions and subroutines that solve the linear equations are as follows:

1. The function GaussJordan implements the Gauss-Jordan method. The function uses local dynamic arrays of indices, with a main for loop to process each column. The main loop contains a nested For loop, which searches for a pivot element and then divides a row by that element. The main loop then reduces the rows, except the row with the pivot element.

2. The function GaussSeidel implements the Gauss-Seidel method. The function normalizes the elements of the matrix A and the vector B by dividing them by the diagonal elements of matrix A. After this step, the function uses a While loop to refine the guesses. Each iteration refines the guesses for the variables and then tests for convergence.

3. The function LUDecomp performs the LU decomposition on matrix A. The function uses a set of nested loops to obtain the largest matrix element, used for scaling all other matrix elements. The function then uses other nested For loops to search for pivot elements and interchange rows. The operations transform the input matrix A into an LU matrix.

4. The function LUBackSubst performs the back substitution to solve for a constant vector. The function uses a For loop for forward substitution, and then uses another For loop for backward substitution.

The Visual Basic Test Program

Let's look at the Visual Basic test program TSMAT.MAK, which tests the various functions that solve linear equations. The program uses a form that has a simple menu system but no controls. Table 1.1 shows the menu structure and the names of the menu items.

The form has the caption "Simultaneous Linear Equations." The menu option Test has three selections to test the three methods for solving linear equations. Each one of these menu selections clears the form and then displays the system of equations being solved and the answers. Thus, you can zoom in on any method by invoking its related menu selection. To compile the test program, you need to include the file MATVECT.BAS in your project file.

Listing 1.2 shows the source code for the test program. The program supplies its own data and then tests the various linear equations methods. Figure 1.1 shows the output of the program Test for each menu selection.

TABLE 1.1 The Structure of the Menu of Program Project TSMAT.MAK

Menu caption	Name
&Test	TesMnu
Gauss-&Jordan Method	GaussJordanMnu
Gauss-&Seidel Method	GaussSeidelMnu
&LU Decomp Method	LUDecompMnu
–	N1
&Exit	ExitMnu

Listing 1.2 The source code for the form of the TSMAT.MAK project.

```
Sub ExitMnu_Click ()
  End
End Sub

Sub GaussJordanMnu_Click ()
  Static A(10, 10) As Double
  Static B(10, 10) As Double
  Dim row As Integer, col As Integer
  Dim numRows As Integer
  Dim numCols As Integer
  Dim myErr As Integer

  numRows = 5
  numCols = 1
  For row = 0 To numRows - 1
   For col = 0 To numRows - 1
```

Listing 1.2 (*Continued*)

```
    A(row, col) = 1
  Next col
 Next row

 A(1, 4) = -1
 A(2, 3) = -1
 A(3, 2) = -1
 A(4, 1) = -1

 B(0, 0) = 15
 B(1, 0) = 5
 B(2, 0) = 7
 B(3, 0) = 9
 B(4, 0) = 11

 Cls
 Print "Testing the Gauss-Jordan Method"
 Print
 For row = 0 To numRows - 1
  For col = 0 To numRows - 1
   If col > 0 Then
   If A(row, col) >= 0 Then
    Print Format$(A(row, col), " +##");
   Else
    Print Format$(Abs(A(row, col)), " -##");
   End If
   Else
    Print Format$(A(row, col), "##");
   End If
    Print Format$(col + 1, " X#");
  Next col
  Print Format$(B(row, 0), " = ##")
 Next row
 myErr = GaussJordan(A(), B(), numRows, numCols)
 If myErr = matErr_None Then
  Print
  Print "Solution vector is:"
  Print
  For row = 0 To numRows - 1
   Print Format$(row + 1, "X(#) = ");
   Print Format$(B(row, 0), "##.###")
  Next row
 Else
  Print "Error in solving equations"
  Print
 End If
End Sub

Sub GaussSeidelMnu_Click ()
 Static A(10, 10) As Double
 Static B2(10) As Double
 Static X(10) As Double
 Dim row As Integer, col As Integer
 Dim numRows As Integer
 Dim myErr As Integer

 A(0, 0) = 10
 A(0, 1) = 1
 A(0, 2) = 1

 A(1, 0) = 1
 A(1, 1) = 10
 A(1, 2) = 1
```

Listing 1.2 (*Continued*)

```
A(2, 0) = 1
A(2, 1) = 1
A(2, 2) = 10

B2(0) = 12
B2(1) = 12
B2(2) = 12

X(0) = 0
X(1) = 0
X(2) = 0

numRows = 3
Cls
Print "Testing the Gauss-Seidel Method"
Print
For row = 0 To numRows - 1
 For col = 0 To numRows - 1
  If col > 0 Then
  If A(row, col) >= 0 Then
   Print Format$(A(row, col), " +##");
  Else
   Print Format$(Abs(A(row, col)), " -##");
  End If
  Else
  Print Format$(A(row, col), "##");
  End If
  Print Format$(col + 1, " X#");
 Next col
 Print Format$(B2(row), " = ##")
Next row

myErr = GaussSeidel(A(), B2(), X(), numRows, 50, .01, .0001)

 If myErr = matErr_None Then
  Print
  Print "Solution vector is:"
  Print
  For row = 0 To numRows - 1
   Print Format$(row + 1, "X(#) = ");
   Print Format$(X(row), "##.###")
  Next row
 Else
  Print "Error in solving equations"
 End If
End Sub

Sub LUDecompMnu_Click ()
 Static A(10, 10) As Double
 Static B2(10) As Double
 Static Index(10) As Integer
 Dim row As Integer, col As Integer
 Dim numRows As Integer
 Dim dFlag As Integer
 Dim myErr As Integer
```

Listing 1.2 *(Continued)*

```
numRows = 5
For row = 0 To numRows - 1
 For col = 0 To numRows - 1
  A(row, col) = 1
 Next col
Next row

A(1, 4) = -1
A(2, 3) = -1
A(3, 2) = -1
A(4, 1) = -1

B2(0) = 15
B2(1) = 5
B2(2) = 7
B2(3) = 9
B2(4) = 11

Cls
Print "Testing the LU Decomposition Method"
Print
For row = 0 To numRows - 1
 For col = 0 To numRows - 1
  If col > 0 Then
   If A(row, col) >= 0 Then
    Print Format$(A(row, col), " +##");
   Else
    Print Format$(Abs(A(row, col)), " -##");
   End If
  Else
   Print Format$(A(row, col), "##");
  End If
  Print Format$(col + 1, " X#");
 Next col
 Print Format$(B2(row), " = ##")
Next row

myErr = LUDecomp(A(), Index(), numRows, dFlag)
If myErr = matErr_None Then
 LUBackSubst A(), Index(), numRows, B2()
 Print
 Print "Solution vector is:"
 Print
 For row = 0 To numRows - 1
  Print Format$(row + 1, "X(#) = ");
  Print Format$(B2(row), "##.###")
 Next row
Else
 Print "Error in solving equations"
End If
End Sub
```

```
Testing the Gauss-Jordan method

1 X1 + 1 X2 + 1 X3 + 1 X4 + 1 X5 = 15
1 X1 + 1 X2 + 1 X3 + 1 X4 - 1 X5 = 5
1 X1 + 1 X2 + 1 X3 - 1 X4 + 1 X5 = 7
1 X1 + 1 X2 - 1 X3 + 1 X4 + 1 X5 = 9
1 X1 - 1 X2 + 1 X3 + 1 X4 + 1 X5 = 11

Solution vector is:
X[1] = 1
X[2] = 2
X[3] = 3
X[4] = 4
X[5] = 5


Testing the Gauss-Seidel method

10 X1 + 1 X2 + 1 X3 = 12
1 X1 + 10 X2 + 1 X3 = 12
1 X1 + 1 X2 + 10 X3 = 12

Solution vector is:
X[1] = 1
X[2] = 1
X[3] = 1


Testing the LU decomposition method

1 X1 + 1 X2 + 1 X3 + 1 X4 + 1 X5 = 15
1 X1 + 1 X2 + 1 X3 + 1 X4 - 1 X5 = 5
1 X1 + 1 X2 + 1 X3 - 1 X4 + 1 X5 = 7
1 X1 + 1 X2 - 1 X3 + 1 X4 + 1 X5 = 9
1 X1 - 1 X2 + 1 X3 + 1 X4 + 1 X5 = 11

Solution vector is:

X[1] = 1
X[2] = 2
X[3] = 3
X[4] = 4
X[5] = 5
```

Figure 1.1 The output of program Test.

Chapter 2

Solving Nonlinear Equations

This chapter looks at a group of popular algorithms involved in solving for the roots of single and multiple nonlinear equations. Most the methods presented in this chapter solve for a single real root of a single nonlinear function. The chapter also includes methods that solve for the multiple roots of a nonlinear function and a polynomial, as well as those that solve for the real roots of multiple nonlinear equations. The chapter discusses the following topics:

- The bisection method
- Newton's method
- The Richmond method
- The combined method
- Brent's method
- The deflating polynomial method
- The Lin-Bairstow method
- Newton's method for solving two nonlinear functions
- Newton's method for solving multiple nonlinear functions

Of the various methods discussed here, the simplest case is the one that finds the single root of a single nonlinear function. The following equation represents the general form of such a function:

$$f(x) = 0 \qquad\qquad (2.1)$$

Equation 2.1 represents a single nonlinear function with one variable. The first set of methods presented in this chapter solves for the value or values of x that conform to equation 2.1. The implementation of the Visual Basic functions in this chapter allow for function f(x) to have an array of fixed parameters. Thus, the Visual Basic functions solve for the following form of function f:

$$f(x, p) = 0 \qquad (2.2)$$

The letter p represents an array of one or more parameters that remain fixed for a particular case, yet vary from one case to another. This approach allows you to use the Visual Basic functions to solve a wide range of mathematical functions, including polynomials.

The chapter also presents dedicated methods that solve for the roots of polynomials. In addition, the chapter presents methods that solve for two or more nonlinear equations. The method that solves for the two roots of two nonlinear equations uses the following forms:

$$F(x, y) = 0 \qquad (2.3)$$

$$G(x, y) = 0 \qquad (2.4)$$

The method solves for the values of variables x and y that fulfill equations 2.3 and 2.4.

The Bisection Method

The bisection method solves for the root of a function f(x) by selecting an initial interval that contains the root and then narrowing that interval down to the desired tolerance level. Here is the algorithm for the bisection method:

Given:

- The interval [A, B], which contains the root of function f(x)
- The tolerance T, which is used to specify the width of the final root-containing interval

Algorithm:

1. Repeat the following steps until $|A - B| < T$:
   1.1. Set guess $G = \dfrac{(A + B)}{2}$
   1.2. If $f(G) * f(B) > 0$, set B = G, otherwise set A = G

2. Return the guess as G or as $\dfrac{(A + B)}{2}$

This algorithm shows the simplicity of the bisection method. The bisection method is a slow converging method, but it is not affected by the slope of the function f(x), as is the case with many of the other methods presented in this chapter. Consequently, the bisection method is slow, but reliable. Figure 2.1 depicts the guesses for the root in the bisection method.

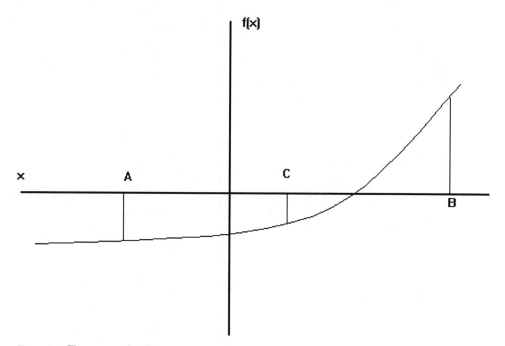

Figure 2.1 The guesses for the root in the bisection method.

Newton's Method

The most popular method for solving for a single root of a single function is Newton's method. This method is, in general, quite suitable for its convergence and computation efforts. Here is the algorithm for Newton's method:

Given:

- The initial guess x for the root of function f(x)
- The tolerance T, which specifies the minimum guess refinement which ends the iteration

Algorithm:

1. Calculate guess refinement $D = \dfrac{f'(x)}{f(x)}$

   where f'(x) is the first derivative of function f(x)

2. Refine guess x by setting x = x – D

3. If |D| > T, then resume at step 1; otherwise, return x as the refined guess for the root

   Newton's method requires the calculation (or estimation) of the first derivative of the function. Figure 2.2 shows the successive guesses for the root in Newton's

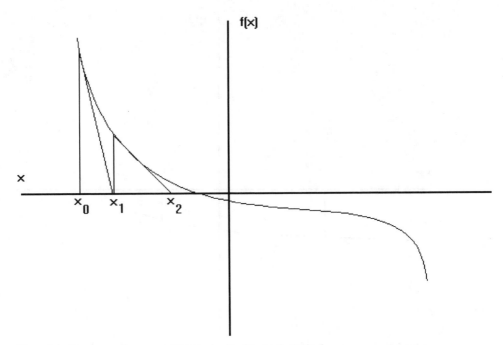

Figure 2.2 The successive guesses for the root in Newton's method.

method. This algorithm assumes that you can calculate the derivative without extensive computational overhead.

Here is the algorithm for the version of Newton's method that estimates the derivative using the function f(x):

Given:

- The initial guess x for the root of function f(x)
- The tolerance T, which specifies the minimum guess refinement which ends the iteration

Algorithm:

1. If |x| > 1, set h = 0.01 * x; otherwise, set h = 0.01

2. Calculate guess refinement $D = \dfrac{(f(x+h) - f(x-h))}{(2 * h * f(x))}$

3. Refine guess x by setting x = x − D

4. If |D| > T, then resume at step 1; otherwise, return x as the refined guess for the root

This algorithm uses a small increment for the current guess to estimate the slope of the function at the guess.

The Richmond Method

The Richmond method has a higher convergence order than Newton's method. However, this advantage does come at a price—the Richmond method requires the first and second derivatives of the function f(x). As with Newton's method, you can implement the algorithm for the Richmond method using either a direct evaluation of the function's derivatives or numerical approximations of the function's derivatives. Here is the algorithm for the Richmond method:

Given:

- The initial guess x for the root of function f(x)
- The tolerance T, which specifies the minimum guess refinement which ends the iteration.

Algorithm:

1. Calculate guess refinement $D = \dfrac{f(x) * f'(x)}{(f'(x)^2 - 0.5 * f(x) * f''(x))}$

    where f'(x) and f''(x) are the first and second derivatives of function f(x), respectively

2. Refine guess x by setting x = x − D

3. If |D| > T, then resume at step 1; otherwise, return x as the refined guess for the root

   Here is the algorithm that uses the approximations to the first and second derivatives:

Given:

- The initial guess x for the root of function f(x)
- The tolerance T, which specifies the minimum guess refinement which ends the iteration

Algorithm:

1. If |x| > 1, set h = 0.01 * x; otherwise, set h = 0.01.

2. Set $fd1 = \dfrac{(f(x + h) - f(x - h))}{2 * h}$

3. Set $fd2 = \dfrac{(f(x + h) - 2 * f(x) + f(x - h))}{h^2}$

4. Calculate guess refinement $D = \dfrac{f(x) * fd1}{(fd1^2 - 0.5 * f(x) * fd2)}$

5. Refine guess x by setting x = x − D

6. If |D| > T, then resume at step 1; otherwise, return x as the refined guess for the root

   Use these algorithms if calculating the derivatives is more time-consuming than calculating the values of the function at x, x + h, and x − h.

The Combined Method

It is possible to devise a method that combines different basic root-seeking algorithms. For example, you can combine the bisection and the Newton methods to take advantage of the strong points of both methods. Such a combination works well when you are dealing with a function that has a small slope in the vicinity of the root. Here is the algorithm for the combined method:

Given:

- The interval [A, B], which contains the root of function f(x)
- Initial guess x
- The tolerance T

Algorithm:

1. Refine guess x using Newton's method
2. If refined guess is not in interval [A, B], perform steps 3 and 4
3. Calculate a new refined guess using the bisection method and the interval [A, B]
4. Shrink the interval [A, B]
5. If guess refinement > T, resume at step 1; otherwise, return x as the sought root

I developed this algorithm in 1980. While researching the algorithms for this book, I found Brent's method, which uses a similar approach.

Newton's Multiroot Method

Newton's method essentially solves for one root. It is possible to incorporate this method (or any other root-solving method) in a more sophisticated algorithm that obtains all of the real roots of a function.

I developed this algorithm using a pseudo-deflation technique. This technique avoids zooming in on roots already obtained by dividing the function f(x) by the product of x minus a previously determined root. Consequently, the modified function steers away from previous roots as they become points in which the modified function possesses infinite values.

Since you are dealing with discontinuities in the modified function, use this method with care and test it before applying it to the kind of functions you want to solve. Another caveat is the fact that the accuracy of the roots suffers as you obtain more roots. If your initial tolerance is small enough, you will most likely obtain roots with acceptable accuracy.

Here is the algorithm for the multiple-root method:

Given:

- The function f(x)
- The maximum number of roots N

- The maximum number of iterations M
- The tolerance factor T
- The array of roots R and the root counter Nroot

Algorithm:

1. Set x = R(0)
2. Set iter = 0
3. Set h = 0.01 * x if |x| > 1; otherwise, set h = 0.01
4. Set f1 = f(x), f2 = f(x + h), and f3 = f(x − h)
5. If Nroot > 1, perform step 7
6. For i = 0 to Nroot − 1, divide f1, f2, and f3 by (x − R(i))
7. Calculate root refinement as $D = \dfrac{2 * h * f1}{(f2 - f3)}$
8. Update root x using x = x − D
9. Increment iter
10. If iter < M and |D| > T, resume at step 1
11. If |D| < T perform the following steps:
    11.1. Store new root using R(NRoot) = x
    11.2. Increment NRoot
    11.3. Update guess x to search for the next root
12. If iter <= M AND Nroot < N, resume at step 1
13. Return the roots in array R

The Deflating Polynomial Method

The algorithms presented so far in this chapter make no assumptions about the form of function f(x). This section handles the case of applying Newton's method to solving the real roots of polynomials with real coefficients.

This deflating polynomial method, also known as the Birge-Vieta method, incorporates Newton's method with a polynomial-deflating step to create the reduced polynomial. In other words, the method solves for the various roots in cycles. The initial cycle solves for a root of the polynomial you specify. The next cycle solves for the root of the polynomial after the reduction of its original polynomial. The subsequent cycles repeat the same tasks.

Here is the algorithm for the Birge-Vieta method:

Given:

- A polynomial of order N and with coefficients A[0] through A[n − 1]
- The maximum iteration limit M

- The initial root guess x
- The array of roots Root

Algorithm:

1. Set iter = 0 and NRoot = 0
2. Repeat the next steps while iter < M and N > 1
3. Increment iter
4. Set z = x
5. Set B[n – 1] = A[n – 1] and C[n – 1] = A[n – 1]
6. For i = n – 2 down to 0, repeat the following:
    6.1. Set B[i] = A[i] + z * B[i + 1]
    6.2. Set C[i] = B[i] + z * C[i + 1]
7. Set B[0] = A[0] + z * B[1]
8. Set D = $\dfrac{B[0]}{C[1]}$
9. Set x = x – D
10. If |D| <= T perform the following:
    10.1. Set iter = 1
    10.2. Decrement N
    10.3. Set Root(Nroot) = x
    10.4. Increment Nroot
    10.5. For i = 0 to N – 1 set A[i] = B[i + 1]
    10.6. If N is equal to 2, perform the following tasks:
        10.6.1. Decrement N
        10.6.2. Set Root(NRoot) = – A[0]
        10.6.3. Increment Nroot
11. Return the results in array Root and counter NRoot

The Lin-Bairstow Method

The deflation polynomial method solves for only the real roots of a polynomial. A method developed by Bairstow and refined by Lin is able to solve for all real and complex roots of a polynomial with real coefficients.

The basic approach works as follows: the method iteratively determines the coefficients of a quadratic polynomial to be factored out of the main polynomial. Then, the method easily calculates the roots of the quadratic polynomial. Next, the method repeats the previous general steps with the deflated polynomials until all the roots are obtained.

Here is the algorithm for the Lin-Bairstow method:

Given:

- A polynomial with order N and the polynomial coefficients A[1] to A[N + 1], where A[1] is the coefficient associated with the term that has the highest power
- The tolerance factor T
- The array of roots R, which has the fields real, image, and .isComplex

Algorithm:

1. If A[1] is not 1, divide coefficients A[N + 1] down to A[1] by A[1]
2. Set root counter Nroot to 0
3. Repeat the next steps until N < 2
   3.1. Set A1 = 1 and B1 = 1
      3.1.1. Set B[0] = 0, D[0] = 0, B[1] = 1, and D[1] = 1
      3.1.2. For i = 2 to N + 1, repeat the following steps:
         3.1.2.1. Set B[i] = A[i] − A1 * B[i − 1] − B1 * B[i − 2]
         3.1.2.2. Set D[i] = B[i] − A1 * D[i − 1] − B1 * D[i − 2]
      3.1.3. Set D1 = D[N − 1]^2 − (D[N] − B[N]) * D[N − 2]

      3.1.4. Set A2 = $\dfrac{(B[N] * D[N - 1] - B[N + 1] * D[N - 2])}{D1}$

      3.1.5. Set B2 = $\dfrac{(B[N + 1] * D[N - 1] - (D[n] * B[n] - B[N]^2))}{D1}$

      3.1.6. Add A2 to A1
      3.1.7. Add B2 to B1
   3.2. If |A2| > T or |B2| > T, resume at step 3.1.1
   3.3. Set D1 = A1 * A1 − 4 * B1
   3.4. If D1 < 0, then calculate the following:

      3.4.1. Set D2 = $\dfrac{\text{SQRT}(|D1|)}{2}$

      3.4.2. Set D3 = $\dfrac{-A1}{2}$

      3.4.3. Set Root(Nroot).isComplex to true
      3.4.4. Set Root(Nroot).real = D3
      3.4.5. Set Root(Nroot).imag = D2
      3.4.6. Set Root(Nroot + 1).isComplex to true
      3.4.7. Set Root(Nroot + 1).real = D3
      3.4.8. Set Root(Nroot + 1).imag = −D2
   3.5. If D1 >= 0, then calculate the following:
      3.5.1. Set Root(Nroot).isComplex to false

      3.5.2. Set Root(Nroot).real = $\dfrac{(D2 - A1)}{2}$

3.5.3. Set Root(Nroot).imag = 0

3.5.4. Set Root(Nroot + 1).isComplex to false

3.5.5. Set Root(Nroot + 1).real = $\dfrac{-(D2 + A1)}{2}$

3.5.6. Set Root(Nroot + 1).imag = 0

3.6. Add 2 to Nroot

3.7. Decrement N by 2

3.8. If N >= 2, then set A[i] = B[i] for i = 1 to N + 1

4. If N is 1, perform the following tasks:

4.1. Set Root(Nroot).isComples to false

4.2. Set Root(Nroot).real = –B[2]

4.3. Set Root(Nroot).image = 0

Solving Multiple Nonlinear Equations

You can extend Newton's method to solve two or more equations. In the case of three or more equations, you need to use the routines that solve for linear equations. These routines allow you to refine the guesses for the multiple roots.

Here is the algorithms for solving two nonlinear equations $F(x, y) = 0$ and $G(x, y) = 0$:

Given:

- The equations $F(x, y) = 0$ and $G(x, y) = 0$
- The initial guesses x and y
- The tolerance factor T

Algorithm:

1. Set F = F(x, y)

2. Set G = G(x, y)

3. Calculate the increments hx and hy for x and y, respectively

4. Set Fx = approximate derivative of function F(x, y) at x

5. Set Fy = approximate derivative of function F(x, y) at y

6. Set Gx = approximate derivative of function G(x, y) at x

7. Set Gy = approximate derivative of function G(x, y) at y

8. Set Jacobian matrix J = Fx * Gy – Fy * Gx

9. Set refinement for x, diffX = $\dfrac{(F * Gy - G * Fy)}{J}$

10. Set refinement for y, $\mathrm{diffY} = \dfrac{(G * Fx - F * Gx)}{J}$

11. Set x = x − diffX and y = y − diffY

12. If |diffX| > T and |diffY| > T, resume at step 1

13. Return the roots x and y for the functions F(x, y) and G(x, y)

Newton's Method for Multiple Equations

You can expand the last algorithm to solve for the multiple real roots of multiple non-linear equations. Solving the roots of two functions is a special and simple case of solving for the roots of multiple equations. In the case of three or more nonlinear functions, you proceed in using Visual Basic functions that solve for simultaneous linear equations. These functions assist in calculating the refinements for the roots' guesses in each iteration.

Here is the algorithm for Newton's method:

Given:

- N nonlinear functions $F^i(x)$ where x is an array of variables
- The tolerance factor T

Algorithm:

1. Copy the values of array x into array x0
2. Store the values of functions $F^i(x)$ in array F0
3. Calculate the matrix, M, of approximate derivates for each function and for each variable
4. Solve for the simultaneous linear equations M D = F0, where D is the array of guess refinements
5. Update the array x using the values in array D
6. If any absolute value in array D exceeds the value of T, resume at step 1
7. Return the roots in array x

The Visual Basic Source Code

Listing 2.1 is the Visual Basic source code that implements the algorithms for the various root-seeking methods just discussed. The module file is named ROOT.BAS.

☞ Throughout the book, the underscore character is used to split wrapping lines of Visual Basic declarations and statements.

Listing 2.1 The source code for the ROOT.BAS module file.

```
Global Const ROOT_EPS# = 1E-30

Type polyRoot
 real As Double
 imag As Double
 isComplex As Integer
End Type

Function Bisection (low As Double, high As Double, _
 tolerance As Double, root As Double, params() As Double) _
 As Integer

 If MyFx(low, params()) * MyFx(high, params()) > 0 Then
   Bisection = False
   Exit Function
 End If

 Do While Abs(high - low) > tolerance
  ' update guess
  root = (low + high) / 2
  If (MyFx(root, params()) * MyFx(high, params())) > 0 Then
   high = root
  Else
   low = root
  End If
 Loop
 Bisection = True
End Function

Function Brent (low As Double, high As Double, _
 tolerance As Double, root As Double, maxIter As Integer, _
 params() As Double) As Integer

 Const SMALL# = .0000001' epsilon
 Const VERY_SMALL# = .0000000001' near zero
 Dim iter As Integer
 Dim a As Double, b As Double, c As Double
 Dim fa As Double, fb As Double, fc As Double
 Dim d As Double, e As Double, tol As Double
 Dim small1 As Double, small2 As Double, small3 As Double
 Dim p As Double, q As Double, r As Double
 Dim s As Double, xm As Double

 iter = 1
 a = low
 b = high
 c = high
 fa = MyFx(low, params())
 fb = MyFx(high, params())
 fc = fb

 ' check that the guesses contain the root
 If (fa * fb) > 0 Then
   Brent = False ' bad guesses
   Exit Function
 End If

 ' start loop to refine the guess for the root
 Do While iter <= maxIter
  iter = iter + 1
  If (fb * fc) > 0 Then
   c = a
   fc = fa
```

Listing 2.1 (*Continued*)

```
  e = b - a
  d = e
End If
If Abs(fc) < Abs(fb) Then
 a = b
 b = c
 c = a
 fa = fb
 fb = fc
 fc = fa
End If
tol = 2 * SMALL * Abs(b) + tolerance / 2
xm = (c - b) / 2
If (Abs(xm) <= tol) Or (Abs(fb) <= VERY_SMALL) Then
 root = b
 Brent = True
 Exit Function
End If
If (Abs(e) >= tol) And (Abs(fa) > Abs(fb)) Then
 ' perform the inverse quadratic interpolation
 s = fb / fa
 If Abs(a - c) <= VERY_SMALL Then
  p = 2 * xm * s
  q = 1 - s
 Else
  q = fa / fc
  r = fb / fc
  q = s * (2 * xm * q * (q - r) - (b - a) * (r - 1))
  q = (q - 1) * (r - 1) * (s - 1)
 End If
 ' determine if improved guess is inside
 ' the range
 If p > 0 Then q = -q
 p = Abs(p)
 small1 = 3 * xm * q * Abs(tol * q)
 small2 = Abs(e * q)
 If small1 < small2 Then
  small3 = small1
 Else
  small3 = small2
 End If
 If (2 * p) < small3 Then
  ' interpolation successful
  e = d
  d = p / q
 Else
  ' use bisection because the interpolation
  ' did not succeed
  d = xm
  e = d
 End If
Else
 ' use bisection because the range
 ' is slowly decreasing
 d = xm
 e = d
End If
' copy most recent guess to variable a
a = b
fa = fb
' evaluate improved guess for the root
If Abs(d) > tol Then
 b = b + d
```

Listing 2.1 (*Continued*)

```
  Else
   If xm > 0 Then
    b = b + Abs(tol)
   Else
    b = b - Abs(tol)
   End If
  End If
  fb = MyFx(b, params())
 Loop
 Brent = False
End Function

Function Combined (low As Double, high As Double, _
 tolerance As Double, root As Double, maxIter As Integer, _
 params() As Double) As Integer

 Dim iter As Integer
 Dim h As Double, diff As Double

 iter = 0
 Do
   iter = iter + 1
   h = .01 * root
   If Abs(root) < 1 Then h = .01
   ' calculate guess refinement
   diff = 2 * h * MyFx(root, params()) / _
    (MyFx(root + h, params()) - MyFx(root - h, params()))
   root = root - diff
   ' check if Newton's method yields a refined guess
   ' outside the range (low, high)
   If (root < low) Or (root > high) Then
    ' apply Bisection method for this iteration
    root = (low + high) / 2
    If (MyFx(root, params()) * MyFx(high, params())) > 0 Then
     high = root
    Else
     low = root
    End If
   End If
 Loop While (iter <= maxIter) And (Abs(diff) > tolerance)

 If Abs(diff) <= tolerance Then
  Combined = True
 Else
  Combined = False
 End If
End Function

Function DeflatePolyRoots (coeff() As Double, _
 initGuess As Double, roots() As Double, numRoots As Integer, _
 polyOrder As Integer, maxIter As Integer, _
 tolerance As Double) As Integer

 Dim a() As Double
 Dim b() As Double
 Dim c() As Double
 Dim diff As Double
 Dim z As Double, X As Double
 Dim i As Integer, iter As Integer, n As Integer

 iter = 1
 n = polyOrder + 1
 X = initGuess
 numRoots = 0
```

Listing 2.1 (*Continued*)

```
' allocate dynamic coefficients
ReDim a(n)
ReDim b(n)
ReDim c(n)

For i = 0 To n - 1
  a(i) = coeff(i)
Next i

Do While (iter <= maxIter) And (n > 1)
  iter = iter + 1
  z = X
  b(n - 1) = a(n - 1)
  c(n - 1) = a(n - 1)
  For i = n - 2 To 1 Step -1
    b(i) = a(i) + z * b(i + 1)
    c(i) = b(i) + z * c(i + 1)
  Next i
  b(0) = a(0) + z * b(1)
  diff = b(0) / c(1)
  X = X - diff
  If Abs(diff) <= tolerance Then
    iter = 1 ' reset iteration counter
    n = n - 1
    roots(numRoots) = X
    numRoots = numRoots + 1
    ' update deflated roots
    For i = 0 To n - 1
      a(i) = b(i + 1)
      ' get the last root
      If n = 2 Then
        n = n - 1
        roots(numRoots) = -a(0)
        numRoots = numRoots + 1
      End If
    Next i
  End If
Loop
DeflatePolyRoots = True
End Function

Function LBPolyRoots (coeff() As Double, roots() As polyRoot, _
 polyOrder As Integer, tolerance As Double) As Integer
'
' solves for the roots of the following polynomial
'
'   y = coeff(0) + coeff(1) X + coeff(2) X^2 +...+ coeff(n) X^n
'
' Parameters:
'
'  coeff    must be an array with at least polyOrder+1 elements.
'
'  roots    output array of roots
'
'  polyOrder order of polynomial
'
'  tolerance tolerance of solutions
'
'
'
  Const SMALL# = .00000001
  Dim a() As Double
  Dim b() As Double
  Dim c() As Double
```

Listing 2.1 *(Continued)*

```
Dim d() As Double
Dim alfa1 As Double, alfa2 As Double
Dim beta1 As Double, beta2 As Double
Dim delta1 As Double, delta2 As Double, delta3 As Double
Dim i As Integer, j As Integer, k As Integer
Dim count As Integer
Dim n As Integer
Dim n1 As Integer
Dim n2 As Integer

n = polyOrder
n1 = n + 1
n2 = n + 2
' is the coefficient of the highest term zero?
If Abs(coeff(0)) < SMALL Then
 LBPolyRoots = False
 Exit Function
End If

' allocate dynamic coefficients
ReDim a(n1)
ReDim b(n1)
ReDim c(n1)
ReDim d(n1)

For i = 0 To n
  a(n1 - i) = coeff(i)
Next i

' is highest coeff not close to 1?
If Abs(a(1) - 1) > SMALL Then
  ' adjust coefficients because a(1) != 1
  For i = 2 To n1
   a(i) = a(i) / a(1)
  Next i
  a(1) = 1#
End If

' initialize root counter
count = 0
Do
  '
  ' start the main Lin-Bairstow iteration loop
  ' initialize the counter and guesses for the
  ' coefficients of quadratic factor:
  '
  ' p(x) = x^2 + alfa1 * x + beta1
  '

  alfa1 = 1
  beta1 = 1

  Do
   b(0) = 0
   d(0) = 0
   b(1) = 1
   d(1) = 1

   j = 1
   k = 0
   For i = 2 To n1
    b(i) = a(i) - alfa1 * b(j) - beta1 * b(k)
    d(i) = b(i) - alfa1 * d(j) - beta1 * d(k)
```

Listing 2.1 (*Continued*)

```
       j = j + 1
       k = k + 1
    Next i
    j = n - 1
    k = n - 2
    delta1 = d(j) ^ 2 - (d(n) - b(n)) * d(k)
    alfa2 = (b(n) * d(j) - b(n1) * d(k)) / delta1
    beta2 = (b(n1) * d(j) - (d(n) - b(n)) * b(n)) / delta1
    alfa1 = alfa1 + alfa2
    beta1 = beta1 + beta2
   Loop While (Abs(alfa2) > tolerance) Or _
        (Abs(beta2) > tolerance)

   delta1 = alfa1 ^ 2 - 4 * beta1

   If delta1 < 0 Then
    ' imaginary roots
    delta2 = Sqr(Abs(delta1)) / 2
    delta3 = -alfa1 / 2
    For i = 0 To 1
     roots(count + i).isComplex = True
     roots(count + i).real = delta3
     roots(count + i).imag = Sign(i) * delta2
    Next i
   Else
    delta2 = Sqr(delta1)
    ' roots are real
    For i = 0 To 1
     roots(count + i).isComplex = False
     roots(count + i).imag = 0
    Next i
    roots(count).real = (delta2 - alfa1) / 2
    roots(count + 1).real = (delta2 + alfa1) / (-2)
   End If
   ' update root counter
   count = count + 2

   ' reduce polynomial order
   n = n - 2
   n1 = n1 - 2
   n2 = n2 - 2

   ' for n >= 2 calculate coefficients of
   ' the new polynomial
   If n >= 2 Then
    For i = 1 To n1
     a(i) = b(i)
    Next i
   End If
  Loop While n >= 2

  If n = 1 Then ' obtain last single real root
   roots(count).isComplex = False
   roots(count).real = -b(2)
   roots(count).imag = 0
  End If

  LBPolyRoots = True
End Function

Function Newton2Functions (rootX As Double, rootY As Double, _
  tolerance As Double, maxIter As Integer, paramsX() As Double, _
  paramsY() As Double) As Integer
```

Listing 2.1 (*Continued*)

```
Dim Jacob As Double
Dim fx0 As Double, fy0 As Double
Dim hX As Double, hY As Double
Dim diffX As Double, diffY As Double
Dim fxy As Double, fxx As Double
Dim fyy As Double, fyx As Double
Dim X As Double
Dim Y As Double
Dim iter As Integer

X = rootX
Y = rootY
iter = 1
Do
  hX = .01 * rootX
  If Abs(rootX) < 1 Then hX = .01
  hY = .01 * rootY
  If Abs(rootY) < 1 Then hY = .01
  fx0 = My2Fx(X, Y, paramsX())
  fy0 = My2Fy(X, Y, paramsY())
  fxx = (My2Fx(X + hX, Y, paramsX()) - _
     My2Fx(X - hX, Y, paramsX())) / 2 / hX
  fyx = (My2Fy(X + hX, Y, paramsY()) - _
     My2Fy(X - hX, Y, paramsY())) / 2 / hX
  fxy = (My2Fx(X, Y + hY, paramsX()) - _
     My2Fx(X, Y - hY, paramsX())) / 2 / hY
  fyy = (My2Fy(X, Y + hY, paramsY()) - _
     My2Fy(X, Y - hY, paramsY())) / 2 / hY
  Jacob = fxx * fyy - fxy * fyx
  diffX = (fx0 * fyy - fy0 * fxy) / Jacob
  diffY = (fy0 * fxx - fx0 * fyx) / Jacob
  X = X - diffX
  Y = Y - diffY
  iter = iter + 1
Loop While (iter <= maxIter) And ((Abs(diffX) > tolerance) Or _
      (Abs(diffY) > tolerance))

rootX = X
rootY = Y
If (Abs(diffX) <= tolerance) And (Abs(diffY) <= tolerance) Then
 Newton2Functions = True
Else
 Newton2Functions = False
End If
End Function

Function NewtonApprox (root As Double, tolerance As Double, _
 maxIter As Integer, params() As Double) As Integer

 Dim iter As Integer
 Dim h As Double, diff As Double

 iter = 0
 Do
   h = .01 * root
   If Abs(root) < 1 Then h = .01
   ' calculate guess refinement
   diff = 2 * h * MyFx(root, params()) / _
      (MyFx(root + h, params()) - MyFx(root - h, params()))
   ' update guess
   root = root - diff
   iter = iter + 1
 Loop While (iter <= maxIter) And (Abs(diff) > tolerance)
```

Listing 2.1 (*Continued*)

```
  If Abs(diff) <= tolerance Then
   NewtonApprox = True
  Else
   NewtonApprox = False
  End If
End Function

Function NewtonExact (root As Double, tolerance As Double, _
 maxIter As Integer, params() As Double) As Integer

 Dim iter As Integer
 Dim diff As Double

 Do
    ' calculate guess refinement
    diff = MyFx(root, params()) / MyDeriv1(root, params())
    ' update guess
    root = root - diff
    iter = iter + 1
 Loop While (iter <= maxIter) And (Abs(diff) > tolerance)

  If Abs(diff) <= tolerance Then
   NewtonExact = True
  Else
   NewtonExact = False
  End If
End Function

Function NewtonMultiRoots (roots() As Double, _
 numRoots As Integer, maxRoots As Integer, _
 tolerance As Double, maxIter As Integer, params() As Double) _
 As Integer

 Dim iter As Integer, i As Integer
 Dim h As Double, diff As Double
 Dim f1 As Double, f2 As Double, f3 As Double
 Dim root As Double

 numRoots = 0
 root = roots(0)

 Do
  iter = 0
  Do
   h = .01 * root
   If Abs(root) < 1 Then h = .01
   f1 = MyFx(root - h, params())
   f2 = MyFx(root, params())
   f3 = MyFx(root + h, params())
   If numRoots > 0 Then
    For i = 0 To numRoots - 1
     f1 = f1 / (root - h - roots(i))
     f2 = f2 / (root - roots(i))
     f3 = f3 / (root + h - roots(i))
    Next i
   End If
   ' calculate guess refinement
   diff = 2 * h * f2 / (f3 - f1)
   ' update guess
   root = root - diff
   iter = iter + 1
  Loop While (iter <= maxIter) And (Abs(diff) > tolerance)
```

Listing 2.1 (*Continued*)

```
  If Abs(diff) <= tolerance Then
   roots(numRoots) = root
   numRoots = numRoots + 1
   If root < 0 Then
    root = .95 * root
   ElseIf root > 0 Then
    root = 1.05 * root
   Else
    root = .05
   End If
  End If
 Loop While (iter <= maxIter) And (numRoots < maxRoots)

 If numRoots > 0 Then
  NewtonMultiRoots = True
 Else
  NewtonMultiRoots = False
 End If
End Function

Function NewtonSimNLE (X() As Double, numEqns As Integer, _
 tolerance As Double, maxIter As Integer) As Integer

 Dim Xdash() As Double
 Dim Fvector() As Double
 Dim index() As Integer
 Dim Jmat() As Double
 Dim i As Integer, j As Integer
 Dim moreIter As Integer, iter As Integer
 Dim rowSwapFlag As Integer
 Dim h As Double
 Dim dummy As Integer

 iter = 0
 ReDim Jmat(numEqns, numEqns)
 ReDim Xdash(numEqns)
 ReDim Fvector(numEqns)
 ReDim index(numEqns)

 Do
  iter = iter + 1
  ' copy the values of array X into array Xdash
  For i = 0 To numEqns - 1
   Xdash(i) = X(i)
  Next i
  ' calculate the array of function values
  For i = 0 To numEqns - 1
   Fvector(i) = MySNE(X(), i)
  Next i
  ' calculate the Jmat matrix
  For i = 0 To numEqns - 1
   For j = 0 To numEqns - 1
    ' calculate increment in variable number j
    If Abs(X(j)) > 1 Then
     h = .01 * X(j)
    Else
     h = .01
    End If
    Xdash(j) = Xdash(j) + h
    Jmat(i, j) = (MySNE(Xdash(), i) - Fvector(i)) / h
    ' restore incremented value
    Xdash(j) = X(j)
   Next j
```

Listing 2.1 (*Continued*)

```
  Next i
  ' solve for the guess refinement vector
  dummy = LUDecomp(Jmat(), index(), numEqns, rowSwapFlag)
  LUBackSubst Jmat(), index(), numEqns, Fvector()
  ' clear the more-iteration flag
  moreIter = False
  ' update guess and test for convergence
  For i = 0 To numEqns - 1
   X(i) = X(i) - Fvector(i)
   If Abs(Fvector(i)) > tolerance Then moreIter = True
  Next i
  ' check iteration limit
  If moreIter Then
   If iter > maxIter Then
    moreIter = False
   Else
    moreIter = True
   End If
  End If
 Loop While moreIter

 NewtonSimNLE = Not moreIter
End Function

Function RichmondApprox (root As Double, tolerance As Double, _
 maxIter As Integer, params() As Double) As Integer

 Dim iter As Integer
 Dim h As Double, diff As Double
 Dim f1 As Double, f2 As Double, f3 As Double
 Dim fd1 As Double, fd2 As Double

 Do
   h = .01 * root
   If Abs(root) < 1 Then h = .01
   f1 = MyFx(root - h, params())   ' f(x-h)
   f2 = MyFx(root, params())      ' f(x)
   f3 = MyFx(root + h, params())   ' f(x+h)
   fd1 = (f3 - f1) / (2 * h)      ' f'(x)
   fd2 = (f3 - 2 * f2 + f1) / Sqr(h) ' f''(x)
   ' calculate guess refinement
   diff = f1 * fd1 / (fd1 ^ 2 - .5 * f1 * fd2)
   ' update guess
   root = root - diff
   iter = iter + 1
 Loop While (iter <= maxIter) And (Abs(diff) > tolerance)

 If Abs(diff) <= tolerance Then
  RichmondApprox = True
 Else
  RichmondApprox = False
 End If
End Function

Function RichmondExact (root As Double, tolerance As Double, _
 maxIter As Integer, params() As Double) As Integer

 Dim iter As Integer
 Dim diff As Double, f1 As Double
 Dim fd1 As Double, fd2 As Double

 iter = 0
 Do
```

Listing 2.1 (*Continued*)

```
    f1 = MyFx(root, params())
    fd1 = MyDeriv1(root, params())
    fd2 = MyDeriv2(root, params())
    ' calculate guess refinement
    diff = f1 * fd1 / (fd1 ^ 2 - .5 * f1 * fd2)
    ' update guess
    root = root - diff
    iter = iter + 1
  Loop While (iter <= maxIter) And (Abs(diff) > tolerance)

  If Abs(diff) <= tolerance Then
    RichmondExact = True
  Else
    RichmondExact = False
  End If
End Function

Function Sign (X As Integer) As Double
  If X = 0 Then
    Sign = 1
  Else
    Sign = -1
  End If
End Function
```

Listing 2.1 declares the structure polyRoot to support solving the real and complex roots of polynomials using the function LBPolyRoots. The implementation file defines the following Visual Basic functions:

1. The function Bisection implements the bisection method. The function has parameters that allow you to specify the root-containing interval, the tolerance factor, the refined root guess, and the array of parameters mentioned in equation 2.1. The function Bisection solves for the root of the Visual Basic function MyFx.

2. The function NewtonApprox implements the version of Newton's method that approximates the first derivative. The function NewtonApprox has parameters that specify the initial guess for the root (and also report the final guess refinement), the tolerance factor, the maximum number of iterations, and the parameters of the solved function. The function NewtonApprox solves for the root of the Visual Basic function MyFx.

3. The function NewtonExact implements the version of Newton's method that uses the derivative of the solved function. The function NewtonExact has parameters that specify the initial guess for the root (and also report the final guess refinement), the tolerance factor, the maximum number of iterations, and the parameters of the solved function. The function NewtonExact solves for the root of the Visual Basic function MyFx and also uses the Visual Basic function MyDeriv1 to provide the values for the first derivative.

4. The function RichmondApprox implements the version of Richmond's method that approximates the first and second derivatives. The function RichmondApprox has parameters that specify the initial guess for the root (and also report the final guess refinement), the tolerance factor, the maximum number of itera-

tions, and the parameters of the solved function. The function RichmondApprox solves for the root of the Visual Basic function MyFx.

5. The function RichmondExact implements the version of Richmond's method that uses the derivatives of the solved function. The function RichmondExact has parameters that specify the initial guess for the root (and also report the final guess refinement), the tolerance factor, the maximum number of iterations, and the parameters of the solved function. The function RichmondExact solves for the root of the Visual Basic function MyFx and also uses the Visual Basic functions MyDeriv1 and MyDeriv2 to provide the values for the first and second derivatives, respectively.

6. The function Combined impelements the combined method. The function has parameters that specify the root-containing range, the initial guess for the root (and also report the final guess refinement), the tolerance factor, the maximum number of iterations, and the parameters of the solved function. The function Combined solves for the root of the Visual Basic function MyFx.

7. The function Brent impelements Brent's method. The function has parameters that specify the root-containing range, the initial guess for the root (and also report the final guess refinement), the tolerance factor, the maximum number of iterations, and the parameters of the solved function. The function Brent solves for the root of the Visual Basic function MyFx.

8. The function NewtonMultiRoots implements the method that uses Newton's algorithm to solve for the multiple roots of a function. The function has parameters that report the calculated roots, the number of roots found, the maximum roots to find, the tolerance factor, the maximum number of iterations, and the parameters of the solved function. The function NewtonMultiRoots solves for the root of the Visual Basic function MyFx.

9. The functionDeflatePolyRoots implements the deflating polynomial method (the Birge-Vieta method). The function has parameters that pass the roots of the polynomial, the initial guess, the array of roots, the number of roots, the polynomial order, the maximum iterations, and the tolerance factor.

10. The function LBPolyRoots implements the Lin-Bairstow method. The function has parameter that include the array of polynomial coefficient, the array of polyRoot structures (which report the results back to the function caller), the polynomial order, and the tolerance factor. The arrays of the coefficients and solved roots must have a number of elements that exceeds the value of the polynomial order by at least one.

11. The function Newton2Functions solves for the two roots of two nonlinear equations, using a special version of Newton's method. The function has parameters that supply the initial guess for the variables x and y (and also report the latest guess refinement), the tolerance factor, the maximum number of iterations, the parameters for function F(x, y), and the parameters for function G(x, y). The function Newton2Functions solves for the roots of the Visual Basic functions My2Fx and My2Fy.

12. The function NewtonSimNLE implements the method that solves for multiple nonlinear equations. The function has parameters that supply the array of initial guesses (and obtain the sought roots), the number of nonlinear equations, the tolerance factor, and the maximum number of iterations. The function NewtonSimNLE solves for the root of the Visual Basic function MySNE.

All of these Visual Basic functions return a true or false value to reflect the success or failure of their operations.

Listing 2.2 The source code for the module MYROOT.BAS.

```
Function My2Fx (X As Double, Y As Double, params() As Double) _
 As Double
 My2Fx = X * X + Y * Y + params(0)
End Function

Function My2Fy (X As Double, Y As Double, params() As Double) _
 As Double
 My2Fy = X * X - Y * Y + params(0)
End Function

Function MyDeriv1 (X As Double, params() As Double) As Double
 MyDeriv1 = Exp(X) - 2 * params(0) * X
End Function

Function MyDeriv2 (X As Double, params() As Double) As Double
 MyDeriv2 = Exp(X) - 6
End Function

Function MyFx (X As Double, params() As Double) As Double
 MyFx = Exp(X) - params(0) * X ^ 2
End Function

Function MySNE (X() As Double, index As Integer) As Double
 Select Case index
  Case 0
   MySNE = X(1) * Exp(X(0)) - 2

  Case 1
   MySNE = X(0) ^ 2 + X(1) - 4

  Case Else
   MySNE = 1
 End Select
End Function
```

Listing 2.2 contains the following special user functions:

1. The function MyFx represents the single nonlinear function whose root you want to solve.

2. The function MyDeriv1 represents the first derivative of the single nonlinear function whose root you want to solve.

3. The function MyDeriv2 represents the second derivatived of the single nonlinear function whose root you want to solve.

4. The functions My2Fx and My2Fy represent the set of two nonlinear functions whose roots you want to solve.

5. The function MySNE represents the set of simultaneous nonlinear functions (all contained with this Visual Basic function) whose roots you want to solve. The first parameter of this function is the array of guesses. The second parameter is an index for the mathematical function whose code is contained inside function MySNE. The function uses a Select statement to examine the value of the index parameter and determine which mathematical equation to evaluate. Each Case label returns the value of a specific mathematical equation. You can extend this Visual Basic function by adding more Case labels. The number of Case labels must be greater than or equal to the number of nonlinear functions being solved.

The Visual Basic Test Program

Let's look at a test program that applies the root-seeking functions defined in Listing 2.1. Listing 2.3 shows the source code for a form associated with the program project TSROOT.MAK. To compile the test program, you need to include the files ROOT .BAS, MYROOT.BAS, and MATVECT.BAS in the project file.

The project uses a form that has a simple menu system but no controls. Table 2.1 shows the menu structure and the names of the menu items. The form has the caption "Root Finding." The menu option Test has three selections to test the various methods for solving nonlinear equations. Each one of these menu selections clears the form and then displays the solved equation(s) and the roots. Thus, you can zoom in on any method by invoking its related menu selection. The program supplies its own data and then tests the various root seeking methods.

TABLE 2.1 The Menu System for the TSROOT.MAK Project

Menu caption	Name
&Test	TesMnu
Bisection Method	BisectionMnu
Newton's Approx. Method	NewtonApproxMnu
Newton's Exact Method	NewtonExactMnu
Richmond's Approx. Method	RichmondApproxMnu
Richmond's Exact Method	RichmondExactMnu
Combined Method	CombinedMnu
Brent Method	BrentMnu
Multiple Roots	MultiRootsMnu
Polynomial Deflation Method	PolyDefMnu
Lin-Bairstow Method	LinBairstowMnu
Two Roots	TwoRootsMnu
Simultaneous Nonlinear Equations	SneMnu
–	N1
&Exit	ExitMnu

The program tests the following functions:

1. The function Bisection in solving for the following equation:

$$e^x - 3x^2 = 0 \tag{2.5}$$

The test specifies the root-containing interval of [2, 4].

2. The function NewtonApprox to solve equation 2.5. The test specifies the initial root of 3, the tolerance factor of 10^{-8}, and the maximum number of 50 iterations.

3. The function NewtonExact to solve equation 2.5. The test specifies the initial root of 3, the tolerance factor of 10^{-8}, and the maximum number of 50 iterations.

4. The function RichmondApprox to solve equation 2.5. The test specifies the initial root of 3, the tolerance factor of 10^{-8}, and the maximum number of 50 iterations.

5. The function RichmondExact to solve equation 2.5. The test specifies the initial root of 3, the tolerance factor of 10^{-8}, and the maximum number of 50 iterations.

6. The function Combined to solve equation 2.5. The test specifies the root-containing interval of [2, 4], the tolerance factor of 10^{-8}, and the maximum number of 50 iterations.

7. The function Brent to solve equation 2.5. The test specifies the root-containing interval of [2, 4], the tolerance factor of 10^{-8}, and the maximum number of 50 iterations.

8. The function NewtonMultiRoots to solve the multiple roots of equation 2.5. The test supplies 2 as the initial guess for the first root. The program displays the three roots of equation 2.5.

9. The function DeflatePolyRoots to solve for the real roots of the following polynomial:

$$x^3 - 23x^2 + 62x - 40 = 0 \tag{2.6}$$

The program displays the real roots of equation 2.6, which are 1, 2, and 20.

10. The function LBPolyRoots to solve for the roots of equation 2.6.

11. The function Newton2Functions to solve for the roots of the following equations:

$$x^2 + y^2 - 1 = 0 \tag{2.7}$$

$$x^2 + y^2 + 0.5 = 0 \tag{2.8}$$

The test uses the initial guesses of 1 and 3 for x and y, respectively. The test also specifies a tolerance factor of 10^{-8} and a maximum of 50 iterations.

12. The function NewtonSimNLE to solve for the roots of the following equations:

$$x_1 e^{x_0} - 2 = 0 \tag{2.9}$$

$$x_0^2 + x_1 - 4 = 0 \tag{2.10}$$

The test uses the initial guess of −0.16 and 2.7 for x_0 and x_1, respectively. Figure 2.3 shows the output of the sample program for each menu selection.

```
Bisection method
Solving exp(x) - 3 * x * x = 0
Guess interval is [ 2 , 4 ]
Root = 3.73307902365923


Newton's method (Approx. Derivative)

Solving exp(x) - 3 * x * x = 0
Initial guess is 3
Root = 3.73307902863285


Newton's method (Supplied Derivative)

Solving exp(x) - 3 * x * x = 0
Initial guess is 3
Root = 3.73307902863281


Richmond's method (Approx. Derivative)

Solving exp(x) - 3 * x * x = 0
Initial guess is 3
Root = 3.7707868981676


Richmond's method (Supplied Derivative)

Solving exp(x) - 3 * x * x = 0
Initial guess is 3
Root = 3.73307902863281


Combined method

Solving exp(x) - 3 * x * x = 0
Guess interval is [ 2 , 4 ]
Root = 3.7330902863685


Brent's method

Solving exp(x) - 3 * x * x = 0
Guess interval is [ 2 , 4 ]
Root = 3.73308956604


Newton's method (Multiple Roots)

Solving exp(x) - 3 * x * x = 0
Initial guess is 2
Root # 1 = 0.91000756248871
Root # 2 = -0.458962267536949
Root # 3 = 3.73307902863281


Deflation method (Polynomial Roots)
```

Figure 2.3 The output of the program for various methods of solving nonlinear equations.

```
1 * X^3 + -23 * X^2 + 62 * X + -40 = 0
Root # 1 = 1
Root # 2 = 2
Root # 3 = 20


Lin-Bairstow method (Polynomial Roots)
1 * X^3 + -23 * X^2 + 62 * X + -40 = 0
Root # 1 = 1
Root # 2 = 2
Root # 3 = 20


Newton's method for two equations

Solving x * x + y * y + -1 = 0
Solving x * x - y * y + 0.5 = 0
Initial guess for X is 1
Initial guess for Y is 3
Root X = 0.5
Root Y = 0.866025403784439


Newton's method for multiple equations

Solving for:
f1(X0, X1) = X1 * exp(X0) - 2
f2(X0, X1) = X0^2 + X1 - 4
Initial guess for X0 = -0.16
Initial guess for X1 = 2.7
Roots are:
X0 = -0.599124782373856
X1 = 3.64104949514535
```

Figure 2.3 (*Continued*)

Listing 2.3 The source code for the form of the TSROOT.MAK project.

```
Sub BisectionMnu_Click ()
 Dim low As Double, high As Double
 Dim root As Double
 Dim tolerance As Double
 Static params(10) As Double

 params(0) = 3
 tolerance = .00000001
 low = 2
 high = 4

 Cls
 Print "Bisection method"
 Print
 Print "Solving exp(x) - "; params(0); " * x * x = 0"
 Print "Guess interval is ["; low; ","; high; "]"
 If Bisection(low, high, tolerance, root, params()) Then
  Print "Root = "; root
 Else
  Print "Failed to obtain root"
 End If
End Sub

Sub BrentMnu_Click ()
 Dim low As Double, high As Double
 Dim root As Double
```

Listing 2.3 *(Continued)*

```
Dim tolerance As Double
Dim maxIter As Integer
Static params(10) As Double

params(0) = 3
tolerance = .00000001
low = 2
high = 4
maxIter = 50

Cls
Print "Brent's method"
Print
Print "Solving exp(x) - "; params(0); " * x * x = 0"
Print "Guess interval is ["; low; ","; high; "]"
If Brent(low, high, tolerance, root, maxIter, params()) Then
  Print "Root = "; root
Else
  Print "Failed to obtain root"
End If
End Sub
Sub CombinedMnu_Click ()
 Dim low As Double, high As Double
 Dim root As Double
 Dim tolerance As Double
 Dim maxIter As Integer
 Static params(10) As Double

 params(0) = 3
 tolerance = .00000001
 low = 2
 high = 4
 maxIter = 50

 Cls
 Print "Combined method"
 Print
 Print "Solving exp(x) - "; params(0); " * x * x = 0"
 Print "Guess interval is ["; low; ","; high; "]"
 If Combined(low, high, tolerance, root, maxIter, params()) Then
  Print "Root = "; root
 Else
  Print "Failed to obtain root"
 End If
End Sub

Sub ExitMnu_Click ()
 End
End Sub

Sub LinBairstowMnu_Click ()
 Static coeff(10) As Double
 Static polyRoots(10) As polyRoot
 Dim tolerance As Double
 Dim numRoots As Integer
 Dim maxIter As Integer
 Dim polyOrder As Integer
 Dim i As Integer

 maxIter = 50
 tolerance = .00000001
```

Listing 2.3 (*Continued*)

```
polyOrder = 3
coeff(0) = -40
coeff(1) = 62
coeff(2) = -23
coeff(3) = 1

Cls
Print "Lin-Bairstow method (Polynomial Roots)"
Print
For i = polyOrder To 0 Step -1
 If (i > 1) And (Abs(coeff(i)) > SMALL) Then
  Print coeff(i); " * X^"; i; " + ";
 ElseIf (i = 1) And (Abs(coeff(i)) > SMALL) Then
  Print coeff(i); " * X + ";
 ElseIf (i = 0) And (Abs(coeff(i)) > SMALL) Then
  Print coeff(i);
 End If
Next i
Print " = 0"

 If LBPolyRoots(coeff(), polyRoots(), polyOrder, tolerance) Then
  For i = 0 To polyOrder - 1
   If polyRoots(i).isComplex Then
   Print "Root # "; i + 1; " = "; _
       Format$(polyRoots(i).real, "##"); " +i "; _
       Format$(polyRoots(i).imag, "##")
   Else
   Print "Root # "; i + 1; " = "; _
       Format$(polyRoots(i).real, "##")
   End If
  Next i
 Else
  Print "Error using the Lin-Bairstow method"
 End If
End Sub

Sub MultiRootsMnu_Click ()
 Static params(10) As Double
 Static roots(10) As Double
 Dim tolerance As Double
 Dim numRoots As Integer
 Dim maxIter As Integer

 params(0) = 3
 roots(0) = 2
 maxIter = 50
 tolerance = .00000001

 Cls
 Print "Newton's method (Multiple Roots)"
 Print
 Print "Solving exp(x) - "; params(0); " * x * x = 0"
 Print "Initial guess is"; roots(0)
 If NewtonMultiRoots(roots(), numRoots, 10, tolerance, _
            maxIter, params()) Then
  For i = 0 To numRoots - 1
   Print "Root # "; i + 1; " = "; roots(i)
  Next i
 Else
  Print "Failed to obtain root"
 End If
End Sub
```

Listing 2.3   *(Continued)*

```
Sub NewtonApproxMnu_Click ()
 Dim root As Double
 Dim tolerance As Double
 Dim maxIter As Integer
 Static params(10) As Double

 params(0) = 3
 tolerance = .00000001
 root = 3
 maxIter = 50

 Cls
 Print "Newton's method (Approx. Derivative)"
 Print
 Print "Solving exp(x) - "; params(0); " * x * x = 0"
 Print "Initial guess is"; root
 If NewtonApprox(root, tolerance, maxIter, params()) Then
  Print "Root = "; root
 Else
  Print "Failed to obtain root"
 End If
End Sub

Sub NewtonExactMnu_Click ()
 Dim root As Double
 Dim tolerance As Double
 Dim maxIter As Integer
 Static params(10) As Double

 params(0) = 3
 tolerance = .00000001
 root = 3
 maxIter = 50

 Cls
 Print "Newton's method (Supplied Derivative)"
 Print
 Print "Solving exp(x) - "; params(0); " * x * x = 0"
 Print "Initial guess is"; root
 If NewtonExact(root, tolerance, maxIter, params()) Then
  Print "Root = "; root
 Else
  Print "Failed to obtain root"
 End If
End Sub

Sub PolyDefMnu_Click ()
 Const SMALL# = .00000001
 Static coeff(10) As Double
 Static roots(10) As Double
 Dim tolerance As Double
 Dim polyOrder As Integer
 Dim maxIter As Integer
 Dim numRoots As Integer
 Dim i As Integer

 polyOrder = 3
 coeff(0) = -40
 coeff(1) = 62
 coeff(2) = -23
 coeff(3) = 1
 maxIter = 50
 tolerance = .00000001
```

Listing 2.3 (*Continued*)

```
Cls
Print "Deflation method (Polynomial Roots)"
Print
For i = polyOrder To 0 Step -1
 If (i > 1) And (Abs(coeff(i)) > SMALL) Then
  Print coeff(i); " * X^"; i; " + ";
 ElseIf (i = 1) And (Abs(coeff(i)) > SMALL) Then
  Print coeff(i); " * X + ";
 ElseIf (i = 0) And (Abs(coeff(i)) > SMALL) Then
  Print coeff(i);
 End If
Next i
Print " = 0"
If (DeflatePolyRoots(coeff(), 1.1, roots(), numRoots, _
           polyOrder, maxIter, tolerance)) Then
 For i = 0 To numRoots - 1
  Print "Root # "; i + 1; " = "; Format$(roots(i), "##")
 Next i
Else
 Print "Failed to obtain root"
End If
End Sub

Sub RichmondApproxMnu_Click ()
 Dim root As Double
 Dim tolerance As Double
 Dim maxIter As Integer
 Static params(10) As Double

 params(0) = 3
 tolerance = .00000001
 root = 3
 maxIter = 50

 Cls
 Print "Richmond's method (Approx. Derivative)"
 Print
 Print "Solving exp(x) - "; params(0); " * x * x = 0"
 Print "Initial guess is"; root
 If RichmondApprox(root, tolerance, maxIter, params()) Then
  Print "Root = "; root
 Else
  Print "Failed to obtain root"
 End If

End Sub

Sub RichmondExactMnu_Click ()
 Dim root As Double
 Dim tolerance As Double
 Dim maxIter As Integer
 Static params(10) As Double

 params(0) = 3
 tolerance = .00000001
 root = 3
 maxIter = 50

 Cls
 Print "Richmond's method (Supplied Derivative)"
 Print
 Print "Solving exp(x) - "; params(0); " * x * x = 0"
 Print "Initial guess is"; root
```

Listing 2.3 (*Continued*)

```
   If RichmondExact(root, tolerance, maxIter, params()) Then
    Print "Root = "; root
   Else
    Print "Failed to obtain root"
   End If
End Sub

Sub SneMnu_Click ()
 Static Xguess(10) As Double
 Dim tolerance As Double
 Dim numEqns As Integer
 Dim maxIter As Integer
 Dim i As Integer

 numEqns = 2
 Xguess(0) = -.16
 Xguess(1) = 2.7
 tolerance = .00000001
 maxIter = 30
 Cls
 Print "Newton's method for multiple equations"
 Print
 Print "Solving for:"
 Print "f1(X0, X1) = X1 * exp(X0) - 2"
 Print "f2(X0, X1) = X0^2 + X1 - 4"
 For i = 0 To numEqns - 1
  Print "Initial guess for X"; Format$(i, "0"); " = "; _
   Xguess(i)
 Next i
 If NewtonSimNLE(Xguess(), numEqns, tolerance, maxIter) Then
  Print "Roots are:"
  For i = 0 To numEqns - 1
   Print "X"; Format$(i, "0"); " = "; Xguess(i)
  Next i
 Else
  Print "Failed to find roots"
 End If
End Sub

Sub TwoRootsMnu_Click ()
 Static paramsX(10) As Double
 Static paramsY(0) As Double
 Dim rootX As Double
 Dim rootY As Double
 Dim tolerance As Double
 Dim maxIter As Integer

 paramsX(0) = -1
 paramsY(0) = .5
 rootX = 1
 rootY = 3
 tolerance = .00000001
 maxIter = 50

 Cls
 Print "Newton's method for 2 Equations"
 Print
 Print "Solving x * x + y * y + "; paramsX(0); " = 0"
 Print "Solving x * x - y * y + "; paramsY(0); " = 0"
 Print "Initial guess for X is "; rootX
 Print "Initial guess for Y is "; rootY
 If Newton2Functions(rootX, rootY, tolerance, maxIter, _
  paramsX(), paramsY()) Then
```

Listing 2.3 *(Continued)*

```
  Print "Root X = "; rootX
  Print "Root Y = "; rootY
Else
  Print "Failed to obtain roots"
End If
End Sub
```

3

Interpolation

Tables represent finite values of observed or calculated data based on two or more arrays. Often, you need to calculate, or interpolate if you prefer, values that are not listed in the table but are in the range of data in that table. This chapter discusses popular methods for interpolating the functions in the form of $y = f(x)$. The chapter covers the following methods:

- The Lagrangian interpolation

- The Barycentric interpolation

- The equidistant Barycentric interpolation

- The Newton divided difference method

- The Newton difference method

- The cubic spline method

Most of the interpolation methods are based on the Taylor expansion of a function $f(x)$ about a specific value x_0. Each method manipulates the Taylor expansion differently to yield the interpolation equation.

The Lagrangian Interpolation

The Lagrangian interpolation is among the most popular interpolation methods. However, this method is not the most efficient, since it does not save intermediate results for subsequent calculations. Here is the equation for the Lagrangian interpolation:

$$y = \frac{y_0 \left[(x - x_1) \dots (x - x_n)\right]}{\left[(x_0 - x_1) \dots (x_0 - x_n)\right]} + \frac{y_1 \left[(x - x_0) \dots (x - x_n)\right]}{\left[(x_1 - x_0) \dots (x_1 - x_n)\right]} + \qquad (3.1)$$

$$\frac{y_2\,[(x-x_0)\,\dots\,(x-x_n)]}{[(x_2-x_0)\,\dots\,(x_0-x_n)]} + \dots + \frac{y_n\,[(x-x_0)\,\dots\,(x-x_{n-1})]}{[(x_n-x_0)\,\dots\,(x_n-x_{n-1})]}$$

Figure 3.1 shows a sample case of Lagrangian interpolation. The thin curve is the interpolating polynomial.

Here is the algorithm for the Lagrangian interpolation:

Given:

- The arrays Xarr and Yarr, which contains x and y values
- The number of elements in each of these arrays, N
- The interpolating value of x, Xint

Algorithm:

1. Set Yint = 0
2. Repeat the next steps for I = 0 to N − 1:
   2.1. Set product P = Yarr[I]
   2.2. Repeat the next step for J = 0 to N − 1:
      2.2.1. If I differs from J, then set $P = \dfrac{P * (\text{Xint} - \text{Xarr[J]})}{(\text{Xarr[I]} - \text{Xarr[J]})}$
   2.3. Add P to Yint
3. Return the interpolated value Yint

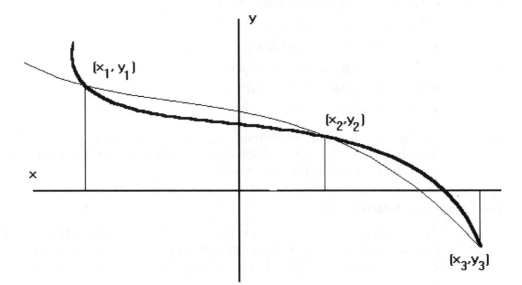

Figure 3.1 A sample case of Lagrangian interpolation. (The thin curve is the interpolating polynomial.)

The Barycentric Interpolation

The algorithm for the Lagrangian method indicates that you need to recalculate all of the terms for each distinct value of x. It is possible to rearrange the equation for the Lagrangian method to define weights that do not depend of the interpolated value of x. Here is the algorithm for this Barycentric method:

Given:

- The arrays Xarr, Yarr, and Wtarr, which represent arrays of x, y, and interpolation weights
- The calculate-weights flag, cwFlag
- The number of elements in the above arrays, N
- The interpolated value of x, Xint

Algorithm:

1. Set sum1 and sum2 to 0
2. If cwFlag is true repeat the next steps for I = 0 to N – 1:
   2.1. Set product P to 1
   2.2. Repeat the following step for J = 0 to N – 1:
      2.2.1. If I and J are not equal, set P = P * (Xarr[I] – Xarr[J])
   2.3. Set Wtarr[I] = $\dfrac{1}{P}$
3. Repeat the next steps for I = 0 to N – 1:
   3.1. Set diff = $\dfrac{wtArr[I]}{(Xint - Xarr[I])}$
   3.2. Set sum1 = sum1 + Yarr[I] * diff
   3.3. Set sum2 = sum2 + diff
4. Set Yint = $\dfrac{sum1}{sum2}$
5. Return Yint as the interpolated value or y

If the array for x contains equidistant data, then the Barycentric algorithm can be simplified:

Given:

- The arrays Xarr and Yarr which represent arrays of x and y
- The number of elements in the arrays, N
- The interpolated value of x, Xint

Algorithm:

1. Set i = 0, w = 1, k = 0, m = N – 1, s = 0, t = 0, x0 = Xarr[0], and h = Xarr[1] – Xarr[0]
2. Repeat the next steps:
   - 2.1. Set diff = Xint – x0
   - 2.2. If diff is very small, set diff to an epsilon value
   - 2.3. Set t = $\dfrac{t + w}{diff}$
   - 2.4. Set s = $\dfrac{s + w * Yarr[i]}{diff}$
   - 2.5. If m is not equal to zero, perform the next steps. Otherwise, resume at step 3:
     - 2.5.1. Set w = w * m
     - 2.5.2. Decrement m
     - 2.5.3. Decrement k
     - 2.5.4. Set w = $\dfrac{w}{k}$
     - 2.5.5. Set x0 = x0 + h
     - 2.5.6. Increment i
     - 2.5.7. Resume at step 2
3. Set Yint = $\dfrac{s}{t}$
4. Return Yint as the interpolated value of y

Newton's Divided Difference Interpolation

Newton's divided difference uses the Taylor expansion in performing the interpolation. The divided differences approximate the various derivatives and allows the method to work with non-equidistant data. The divided differences of the first level approximate the first derivatives. The divided difference of the second level approximate the second derivatives, and so on. Each level of differences has one value less than the preceding level. Here is a sample difference table that also shows the arrays of x and y data:

x_0	y_0	D_0	D_0^2	D_0^3	D_0^4
x_1	y_1	D_1	D_1^2	D_1^3	
x_2	y_2	D_2	D_2^2		
x_3	y_3	D_3			
x_4	y_4				

where

- $D_0 = \dfrac{(y_1 - y_0)}{(x_1 - x_0)}$

- $D_0^2 = \dfrac{(D_1 - D_0)}{(x_1 - x_0)}$

- $D_0^3 = \dfrac{(D_1^2 - D_0^2)}{(x_1 - x_0)}$

and so on. The difference table shown here is an upper triangular matrix. You can store this kind of matrix in a one-dimensional array and save space. The following divided difference algorithm uses such an array to store the table of differences:

Given:

- The arrays Xarr, Yarr, Table, which store the x, y, and difference table data
- The number of elements, N, in each of these arrays
- The build-table flag buildFlag
- The interpolated value of x, Xint

Algorithm:

1. If buildFlag is true, perform the following steps:
   1.1. For I = 0 to N − 1, copy Yarr[I] into Table[I]
   1.2. For I = 0 to N − 2 perform the next step:

      1.2.1. For J = N − 1 down to I + 1, set $\text{Table}[J] = \dfrac{\text{Table}[J-1] - \text{Table}[J]}{(\text{Xarr}[J-1-I] - \text{Xarr}[J])}$

2. Set Yint = Table[N − 1]
3. For I = N − 2 down to 0, set Yint = Table[I] + (Xint − Xarr[I]) ∗ Yint
4. Return Yint as the interpolated value of y

The Newton Difference Method

A special case of the divided difference method emerges when the values of the independent variable are equidistant. This order in the independent variable simplifies the calculations involved. Here is the algorithm for the Newton difference method:

Given:

- The arrays Yarr and Table, which store the y and difference table data
- The number of elements, N, in each of these arrays
- The initial value of x, X0
- The increment in the x values, h
- The build-table flag, buildFlag
- The interpolated value of x, Xint

Algorithm:

1. If buildFlag is true, perform the following steps:
   1.1. For I = 0 to N − 1, copy Yarr[I] into Table[I]
   1.2. For I = 0 to N − 2 perform the next step:

      1.2.1. For J = N − 1 down to I + 1, set $\text{Table}[J] = \dfrac{\text{Table}[J-1] - \text{Table}[J]}{(-h \ast (1 + I))}$

2. Set Yint = Table[N − 1]

3. For I = N − 2 down to 0, set Yint = Table[I] + (Xint − (X0 + I * h)) * Yint

4. Return Yint as the interpolated value of y

The Cubic Spline Interpolation Method

The cubic spline method enjoys a great deal of smoothness for interpolations that involve data that varies significantly. The cubic spline method estimates the second derivatives at the points of reference and then uses these derivatives in the interpolation. Here is the algorithm for calculating the array of second derivatives:

Given:

- The arrays Xarr, Yarr, Deriv, and H, which represent the data for the x, y, second derivatives, and increments
- The number of elements, N, in each of these arrays

Algorithm:

1. For I = 1 to N − 1 perform the following tasks:
   1.1. Set H[I] = Xarr[I] − X[I − 1]
   1.2. Set Deriv[I] = $\dfrac{(Yarr[I] - Yarr[I - 1])}{H[I]}$

2. For I = 1 to N − 2 repeat the following tasks:
   2.1. Set J = I − 1 and K = I + 1
   2.2. Set B[J] = 2
   2.3. Set C[J] = $\dfrac{H[K]}{(H[I] + H[K])}$
   2.4. Set A[J] = 1 − C[J]
   2.5. Set Deriv[J] = $\dfrac{6 * (Deriv[K] - Deriv[I])}{(H[I] + H[K])}$

3. Solve for the second derivatives using the arrays A, B, and C as the arrays of the subdiagonal, diagonal, and superdiagonal elements, respectively. (The array Deriv represents the constants vector.)

Once you obtain the derivatives, you can then apply the following interpolation algorithm:

Given:

- The arrays Xarr, Yarr, Derive, H which represent the data for the x, y, second derivatives, and increments
- The number of elements, N, in each of these arrays
- The interpolated value of x, Xint

Algorithm:

1. Set I = 0
2. Locate the subinterval containing Xint by incrementing the value in I while (Xint < Xarr[I] or Xint > Xarr[I + 1]) and I < (N – 1):
3. Exit if I >= N – 1
4. Set J = I + 1
5. Set Delta1 = Xint – Xarr[I]
6. Set Delta2 = Xarr[J] – Xint
7. Set Yint =

$$\frac{\text{Deriv}[I-1] * \text{delta2}^\wedge 3}{6 * H[J]} + \frac{\text{Deriv}[I] * \text{delta1}^\wedge 3}{6 * H[J]} +$$

$$\left(\frac{\text{Yarr}[J]}{H[J]} - \frac{\text{Deriv}[I] * H[J]}{6} \right) * \text{delta1} +$$

$$\left(\frac{\text{Yarr}[I]}{H[J]} - \frac{\text{Deriv}[I-1] * H[J]}{6} \right) * \text{delta2}$$

8. Return Yint as the interpolated y

The Visual Basic Source Code

Let's look at the Visual Basic source code that implements the interpolation methods given in this chapter. Listing 3.1 shows the source code for the INTERP.BAS module file.

 Throughout the book, the underscore character is used to split wrapping lines of Visual Basic declarations and statements.

Listing 3.1 The source code for the INTERP.BAS module file.

```
Const BAD_RESULT# = -1E-30

Function Barycentric (xarr() As Double, yarr() As Double, _
wt() As Double, calcWtFlag As Integer, n As Integer, _
x As Double) As Double

Dim prod As Double, diff As Double
Dim sum1 As Double, sum2 As Double
Dim i As Integer, j As Integer

sum1 = 0
sum2 = 0

If calcWtFlag Then
  For i = 0 To n - 1
   prod = 1
   For j = 0 To n - 1
   If i <> j Then
    prod = prod * (xarr(i) - xarr(j))
   End If
   Next j
```

Listing 3.1 (*Continued*)

```
   wt(i) = 1 / prod
 Next i
 End If

 ' loop for each term
 For i = 0 To n - 1
  diff = wt(i) / (x - xarr(i))
  sum1 = sum1 + yarr(i) * diff
  sum2 = sum2 + diff
 Next i
 Barycentric = sum1 / sum2
End Function

Function ED_Barycentric (xarr() As Double, yarr() As Double, _
 n As Integer, x As Double) As Double

 Dim h As Double, x0 As Double, diff As Double
 Dim w As Double, s As Double, t As Double
 Dim i As Integer, k As Integer, m As Integer

 i = 0
 w = 1
 k = 0
 m = n - 1
 s = 0
 t = 0
 x0 = xarr(0)
 h = xarr(1) - x0

 Do While True
  diff = x - x0
  If Abs(diff) < INTERP_EPS Then
   diff = INTERP_EPS
  End If
  t = t + w / diff
  s = s + w * yarr(i) / diff
  If m <> 0 Then
   w = w * m
   m = m - 1
   k = k - 1
   w = w / k
   x0 = x0 + h
   i = i + 1
  Else
   Exit Do
  End If
 Loop

 ED_Barycentric = s / t
End Function

Function getDerivatives (xarr() As Double, yarr() As Double, _
 n As Integer, deriv() As Double, h() As Double, _
 tolerance As Double) As Integer

 Dim a() As Double
 Dim b() As Double
 Dim c() As Double
 Dim i As Integer, j As Integer
 Dim k As Integer, result As Integer

 ReDim a(n)
 ReDim b(n)
 ReDim c(n)
```

Listing 3.1 (*Continued*)

```
 For i = 0 To n - 1
  h(i) = 0
  deriv(0) = 0
 Next i

 For i = 1 To n - 1
  j = i - 1
  h(i) = xarr(i) - xarr(j)
  deriv(i) = (yarr(i) - yarr(j)) / h(i)
 Next i

 For i = 1 To n - 2
  j = i - 1
  k = i + 1
  b(j) = 2
  c(j) = h(k) / (h(i) + h(k))
  a(j) = 1 - c(j)
  deriv(j) = 6 * (deriv(k) - deriv(i)) / (h(i) + h(k))
 Next i

 ' solve for the second derivatives
 getDerivatives = tridiagonal(n - 2, a(), b(), c(), deriv(),_
                 tolerance)
End Function

Function Lagrange (xarr() As Double, yarr() As Double, _
 n As Integer, x As Double) As Double

 Dim prod As Double, yint As Double
 Dim i As Integer, j As Integer

 yint = 0

 ' loop for each term
 For i = 0 To n - 1
  ' initialize term with yint[i]
  prod = yarr(i)
  ' build each term
  For j = 0 To n - 1
   If i <> j Then
   prod = prod * (x - xarr(j)) / (xarr(i) - xarr(j))
   End If
  Next j
  yint = yint + prod
 Next i
 Lagrange = yint
End Function

Function NewtonDiff (x0 As Double, h As Double, _
 yarr() As Double, diffTable() As Double, _
 buildMatFlag As Integer, n As Integer, x As Double) As Double

 Dim yint As Double
 Dim i As Integer, j As Integer

 If buildMatFlag Then
  For i = 0 To n - 1
   diffTable(i) = yarr(i)
  Next i

  For i = 0 To n - 2
   For j = n - 1 To i + 1 Step -1
   diffTable(j) = (diffTable(j - 1) - diffTable(j)) / _
          (-h * (1 + i))
```

Listing 3.1 (*Continued*)

```
   Next j
  Next i
End If

 ' initialize interpolated value
 yint = diffTable(n - 1)

 For i = n - 2 To 0 Step -1
  yint = diffTable(i) + (x - (x0 + i * h)) * yint
 Next i

 NewtonDiff = yint
End Function

Function NewtonDivDiff (xarr() As Double, yarr() As Double, _
 diffTable() As Double, buildMatFlag As Integer, n As Integer, _
 x As Double) As Double

 Dim yint As Double
 Dim i As Integer, j As Integer

 If buildMatFlag Then
  For i = 0 To n - 1
   diffTable(i) = yarr(i)
  Next i
  For i = 0 To n - 2
   For j = n - 1 To i + 1 Step -1
    diffTable(j) = (diffTable(j - 1) - diffTable(j)) / _
          (xarr(j - 1 - i) - xarr(j))
   Next j
  Next i
 End If

 ' initialize interpolated value
 yint = diffTable(n - 1)

 For i = n - 2 To 0 Step -1
  yint = diffTable(i) + (x - xarr(i)) * yint
 Next i

 NewtonDivDiff = yint
End Function

Function Spline (xarr() As Double, yarr() As Double, _
 deriv() As Double, h() As Double, n As Integer, x As Double) _
 As Double

 Dim j As Integer, i As Integer
 Dim delta1 As Double, delta2 As Double
 Dim y As Double

 i = 0
 ' locate subinterval containing x
 Do While ((x < xarr(i)) Or (x > xarr(i + 1))) And (i < (n - 1))
  i = i + 1
 Loop

 If (i < 1) Or (i >= (n - 1)) Then
  Spline = BAD_RESULT
  Exit Function
 End If

 j = i + 1
 delta1 = x - xarr(i)
 delta2 = xarr(j) - x
```

Listing 3.1 (*Continued*)

```
y = (deriv(i - 1) * delta2 ^ 3) / (6 * h(j)) + (deriv(i) * _
    delta1 ^ 3) / (6 * h(j)) + (yarr(j) / h(j) - deriv(i) _
    * h(j) / 6) * delta1 + (yarr(i) / h(j) - deriv(i - 1) _
    * h(j) / 6) * delta2

Spline = y
End Function

Private Function tridiagonal (n As Integer, a() As Double, _
b() As Double, c() As Double, d() As Double, _
epsilon As Double) As Integer

Dim isSingular As Integer
Dim i As Integer

isSingular = b(0) < epsilon

' carry out LU factorization
For i = 1 To n - 1
  If isSingular Then Exit For
  a(i) = a(i) / b(i - 1)
  b(i) = b(i) - a(i) * c(i - 1)
  ' determine if diagonal element is too small
  isSingular = b(i) < epsilon
  d(i) = d(i) - a(i) * d(i - 1)
Next i

If Not isSingular Then
  ' carry out backward substitution
  d(n - 1) = d(n - 1) / b(n - 1)
  For i = n - 2 To 0 Step -1
    d(i) = (d(i) - c(i) * d(i + 1)) / b(i)
  Next i
  tridiagonal = True
Else
  tridiagonal = False
End If
End Function
```

Listing 3.1 declares the following Visual Basic functions:

1. The function Lagrange performs the Lagrangian interpolation. The function returns the interpolated value of y and has parameters that represent the arrays of x and y, the number of elements in each of these arrays, and the interpolated value of x.

2. The function Barycentric performs the Barycentric interpolation. The function returns the interpolated value of y and has parameters that represent the arrays of x, y, and weights, the number of elements in each of these arrays, and the interpolated value of x.

3. The ED_Barycentric function performs the equidistant Barycentric interpolation. The function returns the interpolated value of y and has parameters that represent the arrays of x and y, the number of elements in each of these arrays, and the interpolated value of x.

4. The NewtonDivDiff function performs the divided difference interpolation. The function returns the interpolated value of y and has parameters that represent the arrays of x, y, and the difference table; the build-table flag; the number of elements in each of these arrays; and the interpolated value of x.

5. The NewtonDiff function performs the difference interpolation. The function returns the interpolated value of y and has parameters that represent the first value of x, the increment in x, the array of y, the array of the difference table, the build-table flag, the number of elements in each array, and the interpolated value of x.

6. The getDerivatives function obtains the second derivatives used by the function Spline. The function returns an integer value that indicates its success or failure. The parameters of the functions represent the arrays x and y, the number of elements in each of these arrays, the array of derivatives, the array of differences, and the tolerance factor.

7. The function tridiagonal solves the tridiagonal matrix created in function get-Derivatives. The function has parameters that represent the number of linear equations; the subdiagonal, diagonal, and superdiagonal elements; the constants vector; and the tolerance factor.

8. The function Spline performs the cubic spline interpolation. The function returns the interpolated value of y and has parameters that represent the arrays x, y, the derivatives, and the increments; the number of elements in each of these arrays; and the interpolated value of x.

The Visual Basic Test Program

Let's look at a test program that applies the interpolation functions defined in Listing 3.1. Listing 3.2 shows the source code for the project TSINTERP.MAK. To compile the test program, you need to include the file INTERP.BAS in the project file. The project uses a form that has a simple menu system but no controls.

Listing 3.2 The source code for form of project TSINTERP.MAK.

```
Sub BareycentricMnu_Click ()
 Static xarr(5) As Double
 Static yarr(5) As Double
 Static table(5) As Double
 Dim n As Integer
 Dim x As Double, y As Double

 n = 5
 xarr(0) = 1
 xarr(1) = 2
 xarr(2) = 3
 xarr(3) = 4
 xarr(4) = 5
 yarr(0) = 1
 yarr(1) = 4
 yarr(2) = 9
 yarr(3) = 16
 yarr(4) = 25

 Cls
 Print "Barycentric interpolation"
 Print
 x = 2.5
 y = Barycentric(xarr(), yarr(), table(), True, n, x)
 Print "f("; x; ") = "; y
 x = 3.5
 y = Barycentric(xarr(), yarr(), table(), False, n, x)
 Print "f("; x; ") = "; y
End Sub
```

Listing 3.2 *(Continued)*

```
Sub CubicSplineMnu_Click ()
 Static xarr(5) As Double
 Static yarr(5) As Double
 Static deriv(10) As Double
 Static h(10) As Double
 Dim tolerance As Double
 Dim n As Integer
 Dim dummy As Integer
 Dim x As Double, y As Double

 n = 5
 xarr(0) = 1
 xarr(1) = 2
 xarr(2) = 3
 xarr(3) = 4
 xarr(4) = 5
 yarr(0) = 1
 yarr(1) = 4
 yarr(2) = 9
 yarr(3) = 16
 yarr(4) = 25
 tolerance = .000000000000001

 Cls
 Print "Cubic spline interpolation"
 Print
 dummy = getDerivatives(xarr(), yarr(), n, deriv(), h(), _
            tolerance)
 x = 2.5
 y = Spline(xarr(), yarr(), deriv(), h(), n, x)
 Print "f("; x; ") = "; y
 x = 3.5
 y = Spline(xarr(), yarr(), deriv(), h(), n, x)
 Print "f("; x; ") = "; y
 x = 1.5
 y = Spline(xarr(), yarr(), deriv(), h(), n, x)
 Print "f("; x; ") = "; y
End Sub

Sub EDBarycentricMnu_Click ()
 Static xarr(5) As Double
 Static yarr(5) As Double
 Dim n As Integer
 Dim x As Double, y As Double

 n = 5
 xarr(0) = 1
 xarr(1) = 2
 xarr(2) = 3
 xarr(3) = 4
 xarr(4) = 5
 yarr(0) = 1
 yarr(1) = 4
 yarr(2) = 9
 yarr(3) = 16
 yarr(4) = 25

 Cls
 Print "Equidistant Barycentric interpolation"
 Print
 x = 2.5
 y = ED_Barycentric(xarr(), yarr(), n, x)
 Print "f("; x; ") = "; y
 x = 3.5
```

Listing 3.2 (*Continued*)

```
 y = ED_Barycentric(xarr(), yarr(), n, x)
 Print "f("; x; ") = "; y

End Sub

Sub ExitMnu_Click ()
 End
End Sub

Sub LagrangianMnu_Click ()
 Static xarr(5) As Double
 Static yarr(5) As Double
 Dim n As Integer
 Dim x As Double, y As Double

 n = 5
 xarr(0) = 1
 xarr(1) = 2
 xarr(2) = 3
 xarr(3) = 4
 xarr(4) = 5
 yarr(0) = 1
 yarr(1) = 4
 yarr(2) = 9
 yarr(3) = 16
 yarr(4) = 25

 Cls
 Print "Lagrange interpolation"
 Print
 x = 2.5
 y = Lagrange(xarr(), yarr(), n, x)
 Print "f("; x; ") = "; y
 x = 3.5
 y = Lagrange(xarr(), yarr(), n, x)
 Print "f("; x; ") = "; y
End Sub

Sub NewtonDiffMnu_Click ()
 Static xarr(5) As Double
 Static yarr(5) As Double
 Static table(5) As Double
 Dim n As Integer
 Dim x As Double, y As Double

 n = 5
 xarr(0) = 1
 xarr(1) = 2
 xarr(2) = 3
 xarr(3) = 4
 xarr(4) = 5
 yarr(0) = 1
 yarr(1) = 4
 yarr(2) = 9
 yarr(3) = 16
 yarr(4) = 25

 Cls
 Print "Newton difference interpolation"
 Print
 x = 2.5
 y = NewtonDiff(xarr(0), xarr(1) - xarr(0), yarr(), table(), _
        True, n, x)
 Print "f("; x; ") = "; y
 x = 3.5
```

Listing 3.2 *(Continued)*

```
 y = NewtonDiff(xarr(0), xarr(1) - xarr(0), yarr(), table(), _
         False, n, x)
 Print "f("; x; ") = "; y
End Sub

Sub NewtonDivDiffMnu_Click ()
 Static xarr(5) As Double
 Static yarr(5) As Double
 Static table(5) As Double
 Dim n As Integer
 Dim x As Double, y As Double

 n = 5
 xarr(0) = 1
 xarr(1) = 2
 xarr(2) = 3
 xarr(3) = 4
 xarr(4) = 5
 yarr(0) = 1
 yarr(1) = 4
 yarr(2) = 9
 yarr(3) = 16
 yarr(4) = 25

 Cls
 Print "Newton divided difference interpolation"
 Print
 x = 2.5
 y = NewtonDivDiff(xarr(), yarr(), table(), True, n, x)
 Print "f("; x; ") = "; y
 x = 3.5
 y = NewtonDivDiff(xarr(), yarr(), table(), False, n, x)
 Print "f("; x; ") = "; y
End Sub
```

Table 3.1 shows the menu structure and the names of the menu items. The form has the caption "Interpolation." The menu option Test has six selections to test the various methods for interpolation. Each one of these menu selections clears the form and then displays interpolated values. Thus, you can zoom in on any method by invoking its related menu selection. The test program supplies its own data and then tests the various interpolation methods.

TABLE 3.1 The Menu System for the TSINTERP.MAK project

Menu caption	Name
&Test	TesMnu
Lagrange Interpolation	LagrangeMnu
Barycentric Interpolation	BarycentricMnu
Equidistant Barycentric Interpolation	EDBarycentricMnu
Newton's Divided Difference Method	NewtonDivDiffMnu
Newton's Difference Method	NewtonDiffMnu
Cubic Spline Interpolation	CubicSplineMnu
–	N1
&Exit	ExitMnu

The program tests the following functions:

1. The function Lagrange at x = 2.5 and 3.5.
2. The function Barycentric at x = 2.5 and 3.5. This test uses the array table to store the table of data for the second call to function Barycentric.
3. The function ED_Barycentric at x = 2.5 and 3.5.
4. The function NewtonDivDiff at x = 2.5 and 3.5. This test uses the array table to store the table of data for the second call to function NewtonDivDiff.
5. The function NewtonDiff at x = 2.5 and 3.5. This test uses the array table to store the table of data for the second call to function NewtonDiff.
6. The function Spline at x = 2.5, 3.5, and 1.5. This test involves calling the function getDerivatives to first obtain the derivatives required for the interpolation. The calls to function Spline also use the arrays deriv and h to store the derivatives and increments.

Figure 3.2 shows the output of the test program for each menu selection.

```
Lagrange interpolation
f( 2.5 ) = 6.25
f( 3.5 ) = 12.25


Barycentric interpolation

f( 2.5 ) = 6.25
f( 3.5 ) = 12.25


Equidistant Barycentric interpolation

f( 2.5 ) = 6.25
f( 3.5 ) = 12.25


Newton divided difference interpolation

f( 2.5 ) = 6.25
f( 3.5 ) = 12.25


Newton difference interpolation

f( 2.5 ) = 6.25
f( 3.5 ) = 12.25


Cubic spline interpolation

f( 2.5 ) = 6.23214285714826
f( 3.5 ) = 12.2321428571429
f( 1.5 ) = -1E+30
```

Figure 3.2 The output of the program for the different methods of interpolation.

4

Numerical Differentiation

This chapter looks at three methods used to approximate the first four derivatives of a function. These methods are

- The forward/backward difference method
- The central difference method
- The extended central difference method

The chapter presents two sets of Visual Basic functions for these three methods. The first set uses arrays of values, whereas the second set uses user-defined Visual Basic functions. *All the methods assume that the values of the independent variable are equidistant.*

Figure 4.1 shows a simple two-point approximation for a slope.

The Forward/Backward Difference Method

The forward/backward difference method uses the following equations to approximate the first four derivatives of function f(x):

$$f'(x) = \frac{(-3f_0 + 4f_1 - f_2)}{2h} \tag{4.1}$$

$$f''(x) = \frac{(2f_0 - 5f_1 + 4f_2 - f_3)}{2h^2} \tag{4.2}$$

$$f'''(x) = \frac{(-5f_0 + 18f_1 - 24f_2 + 14f_3 - 3f_4)}{2h^3} \tag{4.3}$$

$$f^{iv}(x) = \frac{(3f_0 - 14f_1 + 26f_2 - 24f_3 + 11f_4 - 2f_5)}{h^4} \qquad (4.4)$$

The forward/backward method uses six function values, indexed 0 through 5.

The Central Difference Method

The central difference method uses the following equations to approximate the first four derivatives of function $f(x)$:

$$f'(x) = \frac{(-f_{-1} + f_1)}{2h} \qquad (4.5)$$

$$f''(x) = \frac{(f_{-1} - 2f_0 + f_1)}{h^2} \qquad (4.6)$$

$$f'''(x) = \frac{(-f_{-2} + 2f_{-1} - 2f_1 + f_2)}{2h^3} \qquad (4.7)$$

$$f^{(iv)}(x) = \frac{(f_{-2} - 4f_{-1} + 6f_0 - 4f_1 + f_2)}{h^4} \qquad (4.8)$$

The central difference method uses five function values, indexed –2 through 2.

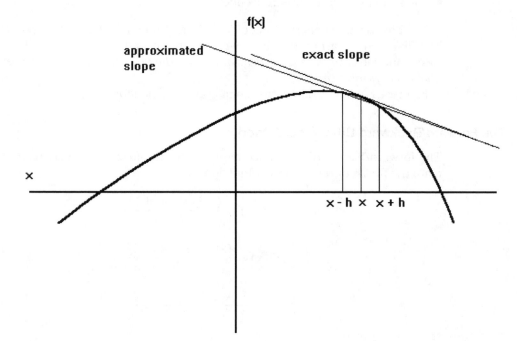

Figure 4.1 A simple two-point approximation for a slope.

The Extended Central Difference Method

The extended central difference method uses the following equations to approximate the first four derivatives of function f(x):

$$f'(x) = \frac{(f_{-2} - 8f_{-1} + 8f_1 - f_2)}{12h} \tag{4.9}$$

$$f''(x) = \frac{(-f_{-2} + 16f_{-1} - 30f_0 + 16f_1 - f_2)}{12h^2} \tag{4.10}$$

$$f'''(x) = \frac{(f_{-3} - 8f_{-2} + 13f_{-1} - 13f_1 + 8f_2 - f_3)}{8h^3} \tag{4.11}$$

$$f^{(iv)}(x) = \frac{(-f_{-3} + 12f_{-2} - 39f_{-1} + 56f_0 - 39f_1 + 12f_2 - f_3)}{6h^4} \tag{4.12}$$

The extended central difference method uses seven function values, indexed –3 through 3.

The Visual Basic Source Code

Let's look at the Visual Basic source code that implements these numerical derivation methods. Listing 4.1 shows the source code for the DERIV.BAS module file. Listing 4.2 contains the source code for the MYDERIV.BAS module file.

 Throughout the book, the underscore character is used to split wrapping lines of Visual Basic declarations and statements.

Listing 4.1 The source code for the DERIV.BAS module file.

```
Global Const DERIV_DEFAULT_INCR# = .01
Global Const DERIV_BAD_RESULT# = -1E+30

Function CDiffDeriv1 (yArr() As Double, incr As Double, _
 n As Integer, y0Idx As Integer) As Double

 ' verify arguments for parameters n and y0Idx
 If n < 3 Or y0Idx < 1 Or y0Idx > (n - 2) Or incr = 0 Then
  CDiffDeriv1 = DERIV_BAD_RESULT
  Exit Function
 End If

 CDiffDeriv1 = (yArr(y0Idx + 1) - yArr(y0Idx - 1)) / 2 / incr
End Function

Function CDiffDeriv2 (yArr() As Double, incr As Double, _
 n As Integer, y0Idx As Integer) As Double

 ' verify arguments for parameters n and y0Idx
 If n < 3 Or y0Idx < 1 Or y0Idx > (n - 2) Or incr = 0 Then
  CDiffDeriv2 = DERIV_BAD_RESULT
  Exit Function
 End If
```

Listing 4.1 (*Continued*)

```
CDiffDeriv2 = (yArr(y0Idx - 1) - 2 * yArr(y0Idx) + _
  yArr(y0Idx + 1)) / incr ^ 2
End Function

Function CDiffDeriv3 (yArr() As Double, incr As Double, _
  n As Integer, y0Idx As Integer) As Double

  ' verify arguments for parameters n and y0Idx
  If n < 5 Or y0Idx < 2 Or y0Idx > (n - 3) Or incr = 0 Then
    CDiffDeriv3 = DERIV_BAD_RESULT
    Exit Function
  End If

  CDiffDeriv3 = (-1 * yArr(y0Idx - 2) + 2 * yArr(y0Idx - 1) - _
    2 * yArr(y0Idx + 1) + yArr(y0Idx + 2)) / 2 / incr ^ 3
End Function

Function CDiffDeriv4 (yArr() As Double, incr As Double, _
  n As Integer, y0Idx As Integer) As Double

  ' verify arguments for parameters n and y0Idx
  If n < 5 Or y0Idx < 2 Or y0Idx > (n - 3) Or incr = 0 Then
    CDiffDeriv4 = DERIV_BAD_RESULT
    Exit Function
  End If

  CDiffDeriv4 = (yArr(y0Idx - 2) - 4 * yArr(y0Idx - 1) + _
    6 * yArr(y0Idx) - 4 * yArr(y0Idx + 1) + _
    yArr(y0Idx + 2)) / incr ^ 4
End Function

Function CUdfDeriv1 (x As Double, incr As Double) As Double

  If incr = 0 Then incr = DERIV_DEFAULT_INCR
  CUdfDeriv1 = (MyFx(x + incr) - MyFx(x - incr)) / 2 / incr
End Function

Function CUdfDeriv2 (x As Double, incr As Double) As Double

  If incr = 0 Then incr = DERIV_DEFAULT_INCR
  CUdfDeriv2 = (MyFx(x - incr) - 2 * MyFx(x) + _
    MyFx(x + incr)) / incr ^ 2
End Function

Function CUdfDeriv3 (x As Double, incr As Double) As Double

  If incr = 0 Then incr = DERIV_DEFAULT_INCR
  CUdfDeriv3 = (-1 * MyFx(x - 2 * incr) + 2 * MyFx(x - incr) - _
    2 * MyFx(x + incr) + MyFx(x + 2 * incr)) / 2 / incr ^ 3
End Function

Function CUdfDeriv4 (x As Double, incr As Double) As Double

  If incr = 0 Then incr = DERIV_DEFAULT_INCR
  CUdfDeriv4 = (MyFx(x - 2 * incr) - 4 * MyFx(x - incr) + _
    6 * MyFx(x) - 4 * MyFx(x + incr) + MyFx(x + 2 * incr)) _
    / incr ^ 4
End Function

Function FBDiffDeriv1 (yArr() As Double, incr As Double, _
  n As Integer, y0Idx As Integer) As Double

  ' verify arguments for parameters n and y0Idx
  If n < 3 Or y0Idx < 0 Or y0Idx > (n - 3) Or incr = 0 Then
```

Listing 4.1 (*Continued*)

```
  FBDiffDeriv1 = DERIV_BAD_RESULT
  Exit Function
 End If

 FBDiffDeriv1 = (-3 * yArr(y0Idx) + 4 * yArr(y0Idx + 1) - _
 yArr(y0Idx + 2)) / 2 / incr
End Function

Function FBDiffDeriv2 (yArr() As Double, incr As Double, _
 n As Integer, y0Idx As Integer) As Double

 ' verify arguments for parameters n and y0Idx
 If n < 4 Or y0Idx < 0 Or y0Idx > (n - 4) Or incr = 0 Then
  FBDiffDeriv2 = DERIV_BAD_RESULT
  Exit Function
 End If

 FBDiffDeriv2 = (2 * yArr(y0Idx) - 5 * yArr(y0Idx + 1) + _
  4 * yArr(y0Idx + 2) - yArr(y0Idx + 3)) / incr ^ 2
End Function

Function FBDiffDeriv3 (yArr() As Double, incr As Double, _
 n As Integer, y0Idx As Integer) As Double

 ' verify arguments for parameters n and y0Idx
 If n < 5 Or y0Idx < 0 Or y0Idx > (n - 5) Or incr = 0 Then
  FBDiffDeriv3 = DERIV_BAD_RESULT
  Exit Function
 End If

 FBDiffDeriv3 = (-5 * yArr(y0Idx) + 18 * yArr(y0Idx + 1) - _
  24 * yArr(y0Idx + 2) + 14 * yArr(y0Idx + 3) - _
  3 * yArr(y0Idx + 4)) / 2 / incr ^ 3
End Function

Function FBDiffDeriv4 (yArr() As Double, incr As Double, _
 n As Integer, y0Idx As Integer) As Double

 ' verify arguments for parameters n and y0Idx
 If n < 6 Or y0Idx < 0 Or y0Idx > (n - 6) Or incr = 0 Then
  FBDiffDeriv4 = DERIV_BAD_RESULT
  Exit Function
 End If

 FBDiffDeriv4 = (3 * yArr(y0Idx) - 14 * yArr(y0Idx + 1) + _
  26 * yArr(y0Idx + 2) - 24 * yArr(y0Idx + 3) + _
  11 * yArr(y0Idx + 4) - 2 * yArr(y0Idx + 5)) / incr ^ 4
End Function

Function FBUdfDeriv1 (x As Double, incr As Double) As Double

 If incr = 0 Then incr = DERIV_DEFAULT_INCR
 FBUdfDeriv1 = (-3 * MyFx(x) + 4 * MyFx(x + incr) - _
 MyFx(x + 2 * incr)) / 2 / incr
End Function

Function FBUdfDeriv2 (x As Double, incr As Double) As Double

 If incr = 0 Then incr = DERIV_DEFAULT_INCR
 FBUdfDeriv2 = (2 * MyFx(x) - 5 * MyFx(x + incr) + _
  4 * MyFx(x + 2 * incr) - MyFx(x + 3 * incr)) / incr ^ 2
End Function

Function FBUdfDeriv3 (x As Double, incr As Double) As Double
```

Listing 4.1 (*Continued*)

```
 If incr = 0 Then incr = DERIV_DEFAULT_INCR
 FBUdfDeriv3 = (-5 * MyFx(x) + 18 * MyFx(x + incr) - _
 24 * MyFx(x + 2 * incr) + 14 * MyFx(x + 3 * incr) - _
 3 * MyFx(x + 4 * incr)) / 2 / incr ^ 3
End Function

Function FBUdfDeriv4 (x As Double, incr As Double) As Double

 If incr = 0 Then incr = DERIV_DEFAULT_INCR
 FBUdfDeriv4 = (3 * MyFx(x) - 14 * MyFx(x + incr) + _
 26 * MyFx(x + 2 * incr) - 24 * MyFx(x + 3 * incr) + _
 11 * MyFx(x + 4 * incr) - 2 * MyFx(x + 5 * incr)) / incr ^ 4
End Function

Function XCDiffDeriv1 (yArr() As Double, incr As Double, _
 n As Integer, y0Idx As Integer) As Double

 ' verify arguments for parameters n and y0Idx
 If n < 5 Or y0Idx < 2 Or y0Idx > (n - 3) Or incr = 0 Then
  XCDiffDeriv1 = DERIV_BAD_RESULT
  Exit Function
 End If

 XCDiffDeriv1 = (yArr(y0Idx - 2) - 8 * yArr(y0Idx - 1) + _
 8 * yArr(y0Idx + 1) - yArr(y0Idx + 2)) / 12 / incr
End Function

Function XCDiffDeriv2 (yArr() As Double, incr As Double, _
 n As Integer, y0Idx As Integer) As Double

 ' verify arguments for parameters n and y0Idx
 If n < 5 Or y0Idx < 2 Or y0Idx > (n - 3) Or incr = 0 Then
  XCDiffDeriv2 = DERIV_BAD_RESULT
  Exit Function
 End If

 XCDiffDeriv2 = (-1 * yArr(y0Idx - 2) + 16 * yArr(y0Idx - 1) - _
 30 * yArr(y0Idx) + 16 * yArr(y0Idx + 1) - _
 yArr(y0Idx + 2)) / 12 / incr ^ 2
End Function

Function XCDiffDeriv3 (yArr() As Double, incr As Double, _
 n As Integer, y0Idx As Integer) As Double

 ' verify arguments for parameters n and y0Idx
 If n < 7 Or y0Idx < 3 Or y0Idx > (n - 4) Or incr = 0 Then
  XCDiffDeriv3 = DERIV_BAD_RESULT
  Exit Function
 End If

 XCDiffDeriv3 = (yArr(y0Idx - 3) - 8 * yArr(y0Idx - 2) + _
 13 * yArr(y0Idx - 1) - 13 * yArr(y0Idx + 1) + _
 8 * yArr(y0Idx + 2) - yArr(y0Idx + 3)) / 8 / incr ^ 3
End Function

Function XCDiffDeriv4 (yArr() As Double, incr As Double, _
 n As Integer, y0Idx As Integer) As Double

 ' verify arguments for parameters n and y0Idx
 If n < 7 Or y0Idx < 3 Or y0Idx > (n - 4) Or incr = 0 Then
  XCDiffDeriv4 = DERIV_BAD_RESULT
  Exit Function
 End If
```

Listing 4.1 (*Continued*)

```
XCDiffDeriv4 = (-1 * yArr(y0Idx - 3) + 12 * yArr(y0Idx - 2) - _
 39 * yArr(y0Idx - 1) + 56 * yArr(y0Idx) - _
 39 * yArr(y0Idx + 1) + 12 * yArr(y0Idx + 2) - _
 yArr(y0Idx + 3)) / 6 / incr ^ 4
End Function

Function XCUdfDeriv1 (x As Double, incr As Double) As Double

 If incr = 0 Then incr = DERIV_DEFAULT_INCR
 XCUdfDeriv1 = (MyFx(x - 2 * incr) - 8 * MyFx(x - incr) + _
 8 * MyFx(x + incr) - MyFx(x + 2 * incr)) / 12 / incr
End Function

Function XCUdfDeriv2 (x As Double, incr As Double) As Double

 If incr = 0 Then incr = DERIV_DEFAULT_INCR
 XCUdfDeriv2 = (-MyFx(x - 2 * incr) + 16 * MyFx(x - incr) - _
 30 * MyFx(x) + 16 * MyFx(x + incr) - _
 MyFx(x + 2 * incr)) / 12 / incr ^ 2
End Function

Function XCUdfDeriv3 (x As Double, incr As Double) As Double

 If incr = 0 Then incr = DERIV_DEFAULT_INCR
 XCUdfDeriv3 = (MyFx(x - 3 * incr) - 8 * MyFx(x - 2 * incr) + _
 13 * MyFx(x - incr) - 13 * MyFx(x + incr) + _
 8 * MyFx(x + 2 * incr) - MyFx(x + 3 * incr)) / 8 / incr ^ 3
End Function

Function XCUdfDeriv4 (x As Double, incr As Double) As Double
 If incr = 0 Then incr = DERIV_DEFAULT_INCR
 XCUdfDeriv4 = (-MyFx(x - 3 * incr) + _
 12 * MyFx(x - 2 * incr) - 39 * MyFx(x - incr) + _
 56 * MyFx(x) - 39 * MyFx(x + incr) + _
 12 * MyFx(x + 2 * incr) - MyFx(x + 3 * incr)) / 6 / incr ^ 4
End Function
```

Listing 4.1 declares the following set of Visual Basic functions:

1. The functions FBDiffDeriv1 through FBDiffDeriv4 implement equations 4.1 through 4.4 using arrays of function values. Each Visual Basic function has parameters that pass the y values, the x increment value, the number of elements in the array y, and the index of array y, which specifies the central element (index 0, as far as the difference equations are concerned).

2. The functions CDiffDeriv1 through CDiffDeriv4 implement equations 4.5 through 4.8 using arrays of function values. Each Visual Basic function has parameters that pass the y values, the x increment value, the number of elements in the array y, and the index of array y, which specifies the central element (index 0, as far as the difference equations are concerned).

3. The functions XCDiffDeriv1 through XCDiffDeriv4 implement equations 4.9 through 4.12 using arrays of function values. Each Visual Basic function has parameters that pass the y values, the x increment value, the number of elements in the array y, and the index of array y, which specifies the central element (index 0, as far as the difference equations are concerned).

4. The functions FBUdfDeriv1 through FBUdfDeriv4 implement equations 4.1 through 4.4 using a user-defined Visual Basic function. Each FBUdfDeriv*X* Visual Basic function has parameters that pass the value of the x and the increment value.

5. The functions CUdfDeriv1 through CUdfDeriv4 implement equations 4.5 through 4.8 using a user-defined Visual Basic function. Each CUdfDeriv*X* Visual Basic function has parameters that pass the value of the x and the increment value.

6. The functions XCUdfDeriv1 through XCUdfDeriv4 implement equations 4.5 through 4.8 using a user-defined Visual Basic function. Each XCUdfDeriv*X* Visual Basic function has parameters that pass the value of the x and the increment value.

Listing 4.2 The source code for the MYDERIV.BAS module file.

```
Function MyFx (X As Double) As Double
 MyFx = X ^ 4 - 1
End Function
```

Listing 4.2 implements the Visual Basic function MyFx. This function is the user-defined function.

The Visual Basic Test Program

Let's look at a test program that applies the interpolation functions defined in Listing 4.1. Listing 4.3 shows the source code for the program project TSDERIV.MAK. To compile the test program, you need to include the files DERIV.BAS and MYDERIV.BAS in your project file. The project uses a form that has a simple menu system but no controls.

Table 4.1 shows the menu structure and the names of the menu items. The form has the caption "Derivation." The menu option Test has three selections to test the various methods for numerical derivation. Each one of these menu selections clears the form, and then displays the values for the first four derivatives. Thus, you can zoom in on any method by invoking its related menu selection.

Listing 4.3 The source code for the form associated with the project TSDERIV.MAK.

```
Const ARRAY_SIZE% = 20

Dim yArr(ARRAY_SIZE) As Double
Dim x As Double
Dim deltaX As Double
Dim i As Integer, y0Idx As Integer


Sub CentralMnu_Click ()
 Cls
 Print "Using Central difference formula"
 Print
 x = 1
 y0Idx = 5
 Print "Using difference values"
 Print "y'("; x; ") = "; Format$(CDiffDeriv1(yArr(), _
```

Listing 4.3 (*Continued*)

```
  deltaX, ARRAY_SIZE, y0Idx), "##.##")
 Print "y''("; x; ") = "; Format$(CDiffDeriv2(yArr(), _
  deltaX, ARRAY_SIZE, y0Idx), "##.##")
 Print "y'''("; x; ") = "; Format$(CDiffDeriv3(yArr(), _
  deltaX, ARRAY_SIZE, y0Idx), "##.##")
 Print "y''''("; x; ") = "; Format$(CDiffDeriv4(yArr(), _
  deltaX, ARRAY_SIZE, y0Idx), "##.##")
 Print
 Print "Using user-defined equation"
 Print "y'("; x; ") = "; Format$(CUdfDeriv1(x, _
  deltaX), "##.##")
 Print "y''("; x; ") = "; Format$(CUdfDeriv2(x, _
  deltaX), "##.##")
 Print "y'''("; x; ") = "; Format$(CUdfDeriv3(x, _
  deltaX), "##.##")
 Print "y''''("; x; ") = "; Format$(CUdfDeriv4(x, _
  deltaX), "##.##")
End Sub


Sub ExitMnu_Click ()
 End
End Sub


Sub ExtCentralMnu_Click ()
 Cls
 Print "Using Extended Central difference formula"
 Print
 x = 1
 y0Idx = 5
 Print "Using difference values"
 Print "y'("; x; ") = "; Format$(XCDiffDeriv1(yArr(), _
  deltaX, ARRAY_SIZE, y0Idx), "##.##")
 Print "y''("; x; ") = "; Format$(XCDiffDeriv2(yArr(), _
  deltaX, ARRAY_SIZE, y0Idx), "##.##")
 Print "y'''("; x; ") = "; Format$(XCDiffDeriv3(yArr(), _
  deltaX, ARRAY_SIZE, y0Idx), "##.##")
 Print "y''''("; x; ") = "; Format$(XCDiffDeriv4(yArr(), _
  deltaX, ARRAY_SIZE, y0Idx), "##.##")
 Print
 Print "Using user-defined equation"
 Print "y'("; x; ") = "; Format$(XCUdfDeriv1(x, deltaX), _
   "##.##")
 Print "y''("; x; ") = "; Format$(XCUdfDeriv2(x, deltaX), _
   "##.##")
 Print "y'''("; x; ") = "; Format$(XCUdfDeriv3(x, deltaX), _
   "##.##")
 Print "y''''("; x; ") = "; Format$(XCUdfDeriv4(x, deltaX), _
   "##.##")
End Sub


Sub FBDiffMnu_Click ()
 Cls
 Print "Using Forward/Backward difference formula"
 Print
 x = 1
 y0Idx = 5
 Print "Using difference values"
 Print "y'("; x; ") = "; Format$(FBDiffDeriv1(yArr(), _
  deltaX, ARRAY_SIZE, y0Idx), "##.##")
 Print "y''("; x; ") = "; Format$(FBDiffDeriv2(yArr(), _
  deltaX, ARRAY_SIZE, y0Idx), "##.##")
 Print "y'''("; x; ") = "; Format$(FBDiffDeriv3(yArr(), _
```

Listing 4.3 (*Continued*)

```
 deltaX, ARRAY_SIZE, y0Idx), "##.##")
 Print "y''''("; x; ") = "; Format$(FBDiffDeriv4(yArr(), _
 deltaX, ARRAY_SIZE, y0Idx), "##.##")
 Print
 Print "Using user-defined equation"
 Print "y('"; x; ") = "; Format$(FBUdfDeriv1(x, deltaX), _
   "##.##")
 Print "y(''"; x; ") = "; Format$(FBUdfDeriv2(x, deltaX), _
   "##.##")
 Print "y('''"; x; ") = "; Format$(FBUdfDeriv3(x, deltaX), _
   "##.##")
 Print "y''''("; x; ") = "; Format$(FBUdfDeriv4(x, deltaX), _
   "##.##")
End Sub

Sub Form_Load ()
 x = .5
 deltaX = .1
 For i = 0 To ARRAY_SIZE - 1
  yArr(i) = MyFx(x)
  x = x + deltaX
 Next i
End Sub
```

TABLE 4.1 The Menu System for the TSDERIV.MAK Project

Menu caption	Name
&Test	TesMnu
Forward/Backward Difference	FBDiffMnu
Central Difference	CentralMnu
Extended Central Difference	ExtCentralMnu
–	N1
&Exit	ExitMnu

The test program supplies its own data and then tests the various interpolation methods. The program creates an array of y values for x ranging from 0.5 to 2.5. The listing also defines the test function as $x^4 - 1$. The program then performs the following tasks:

1. The Forward/Backward Difference menu selection tests the forward/backward difference formula for x = 1 and y0Idx = 5. The program calculates and displays the values of the first four derivatives using functions FBDiffDeriv1 through FBDiffDeriv4 and FBUdfDeriv1 through FBUdffDeriv4.

2. The Central Difference menu selection tests the forward/backward difference formula for x = 1. The program calculates and displays the values of the first four derivatives using functions CDiffDeriv1 through CDiffDeriv4 and CUdfDeriv1 through CUdffDeriv4.

3. The Extended Central Difference menu selection tests the central difference formula for x = 1 and y0Idx = 5. The program calculates and displays the values of the first four derivatives using functions XCDiffDeriv1 through XCDiffDeriv4 and XCUdfDeriv1 through XCUdffDeriv4.

Figure 4.2 shows the output of each menu selection.

```
The output of the Forward/Backward Difference menu selection:

Using difference values
y'(1) = 3.91
y''(1) = 11.78
y'''(1) = 24
y''''(1) = 24

Using user-defined equation
y'(1) = 3.91
y''(1) = 11.78
y'''(1) = 24
y''''(1) = 24


The output of the Central Difference menu selection:

Using difference values
y'(1) = 4.04
y''(1) = 12.02
y'''(1) = 24
y''''(1) = 24

Using user-defined equation
y'(1) = 4.04
y''(1) = 12.02
y'''(1) = 24
y''''(1) = 24


The output of the Extended Central Difference menu selection:

Using difference values
y'(1) = 4
y''(1) = 12
y'''(1) = 24
y''''(1) = 24

Using user-defined equation
y'(1) = 4
y''(1) = 12
y'''(1) = 24
y''''(1) = 24
```

Figure 4.2 The output of each menu selection for the test program.

Chapter 5

Numerical Integration

This chapter looks at popular methods for numerical integration. These method either use arrays of values (with the option of using direct function evaluations) or use function evaluations. The chapter discusses the following methods:

- Simpson's method
- Simpson's alternate method
- Gauss-Legendre quadrature
- Gauss-Laguerre quadrature
- Gauss-Hermite quadrature
- Gauss-Chebyshev quadrature
- Romberg's method

Simpson's Method

The most popular numerical integration method is Simpson's method. In fact, there are several versions of Simpson's method that use points with equidistant values for the independent variable. The most popular Simpson methods are the one-third and three-eights rules.

The equation of the one-third rule requires at least three values of y and grows into odd numbers (5, 7, 9, and so on) of y values. The equation for the Simpson's rule is

$$\int f(x) \, dx = \frac{h \, (y_1 + 4y_2 + y_3)}{3} \tag{5.1}$$

where h is the increment in the independent variable. The general form for this equation is as follows:

$$\int f(x)\,dx = \frac{h\,(y_1 + 4y_2 + 2y_3 + \dots + 2y_{N-2} + 4y_{N-1} + y_N)}{3} \qquad (5.2)$$

Figure 5.1 depicts Simpson's rule. Here is the algorithm for Simpson's one-third rule:

Given:

- The array Yarr with an odd number of elements, N
- The increment h

Algorithm:

1. Set sumEven = 0 and sumOdd = 0
2. For I = 1 to N – 2, add Yarr[I] to sumEven
3. For I = 2 to N – 3, add Yarr[I] to sumOdd
4. Set Area = $\dfrac{h}{3}$ * (Yarr[0] + 4 * sumEven + 2 * sumOdd + Yarr[N – 1]
5. Return Area as the value of the integral

You can use this algorithm with direct function evaluations by replacing Yarr[I] with the expression (X0 + h * I).

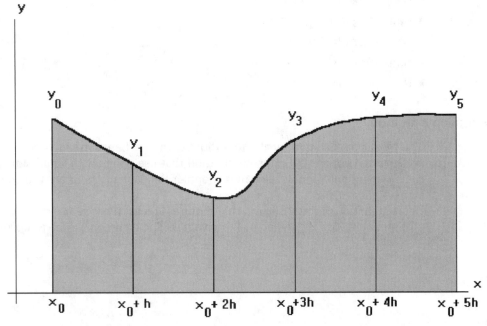

Figure 5.1 Simpson's rule.

Simpson's Alternate Extended Rule

Let's look at another Simpson's rule, which offers better results than the one-third (and three-eights) rules. Here is the equation for the alternate extended rule:

$$\int f(x)\,dx = \tag{5.3}$$

$$\frac{h\,(17f_1 + 59f_2 + 43f_3 + 49f_4 + f_5 + f_6 + \ldots + f_{N-4} + 49f_{N-3} + 43f_{N-2} + 59f_{N-1} + 17f_N)}{48}$$

Here is the algorithm for using the Simpson's alternate extended rule:

Given:

- An array Yarr of y values
- The number of array elements, N
- The increment in the independent variable, h

Algorithm:

1. Set Sum = 0
2. For I = 0 to N − 1 add Yarr[I] to Sum
3. Set Sum = h/48 * (48 * Sum − 31 * Yarr[0] + 11 * Yarr[1] − 5 * Yarr[2] + Yarr[3] − Yarr[N − 4] − 5 * Yarr[N − 3] + 11 * Yarr[N − 2] − 31 * Yarr[N − 1])
4. Return Sum as the numerical integral

You can use this algorithm with direct function evaluations by replacing Yarr[I] with the expression (X0 + h * I).

The Gaussian Quadrature Methods

The Gaussian quadrature methods represent a different way of calculating integrals that enjoy high accuracy. Rather than sampling the function value at regular values of the independent variable, the Gaussian quadrature zoom in on critical-function values. These methods obtain the sought integral based on the product of the critical-function values and special weights.

The Gaussian quadratures can serve observed data if these data can be sampled at the critical values. Another key point to keep in mind regarding Gaussian quadratures is that they have specific limits of integration. You can still use some of these methods to integrate over your own range of x values by mapping the values of your x values (call it the real-world x) with the Gaussian quadrature's x values.

Gauss-Legendre Quadrature

The Gauss-Legendre quadrature uses the Legendre polynomial to obtain the weights and critical points that evaluate an integral in the range of −1 to 1. You can use the Gauss-Legendre quadrature to evaluate the integral for the range [A, B] by using the following equations:

$$\int f(x)\, dx = \frac{(B-A)(c_1\, f(x_1) + c_2\, f(x_2) + \ldots + c_n\, f(x_n))}{2} \qquad (5.4)$$

$$x_i = A + (B-A)(z_i + 1) \qquad (5.5)$$

where z_i is the critical quadrature value in the range -1 to 1, and x_i is the real-world value that corresponds to z_i. Equation 5.4 shows the weights c_1 through c_n. The value of n in equation 5.4 is the order of the Legendre polynomial used to precalculate the values of z_i and c_i.

The Visual Basic source code in this chapter uses values for z_i and c_i based on the Legendre polynomial *order of 6*. This means that the Gauss-Legendre quadrature evaluates an integral with six points. Because integrals may cover a wide range of x values or require enhanced accuracy, here is an algorithm that applies the Gauss-Legendre quadrature over subintervals:

Given:

- The range of x, A, and B for the integration
- The number of subintervals, N, for the range [A, B]
- The function f(x).

Algorithm:

1. Initialize the array Xk to store the critical x values of the quadrature
2. Initialize the array Ak to store the weights
3. Set Area = 0
4. Set $H = \dfrac{(B-A)}{N}$
5. Set xA = A
6. For I = 1 to N, repeat the next steps:
   6.1. Set Sum = 0
   6.2. Set xB = xA + H
   6.3. For J = 0 to 5 repeat the next steps:

   $\qquad$ 6.3.1. Set $X = \left(xA + \dfrac{H}{2}\right) * (Xk[J] + 1)$

   $\qquad$ 6.3.2. Add Ak[J] * f(X) to Sum

   6.4. Add $\left(\dfrac{H}{2} * Sum\right)$ to Area

   6.5. Set xA = xB
7. Return Area as the result of the numerical integration

The Gauss-Laguerre Quadrature

The Gauss-Laguerre quadrature evaluates the integral of $e^x f(x)$ between zero and infinity using the Laguerre polynomials. The method calculates the approximation for the integral by summing the product of quadrature weights and the values of function $f(x)$ at critical points of x. Here is the equation for the Gauss-Laguerre quadrature:

$$\int f(x) \, dx = \Sigma \, A_i \, f(x_i) \tag{5.6}$$

To obtain the integral of a custom function, $g(x)$, you need to write that function as $f(x) = g(x) \, e^{-x}$, and then use $f(x)$ in the integration process. The number of added terms (each term is the product of the weight and function value) depends on the order of the Laguerre polynomial used in approximating the integral.

Here is the algorithm for the Gauss-Laguerre quadrature:

Given:

- The function $f(x)$

Algorithm:

1. Set Sum = 0
2. Initialize the array Xk to store the critical x values of the quadrature
3. Initialize the array Ak to store the weights
4. For I = 0 to 5, add Ak[I] * f(Xk[I]) to Sum
5. Return Sum as the area for $e^x \, f(x)$ between 0 and infinity

The Gauss-Hermite Quadrature

The Gauss-Hermite quadrature evaluates the integral of $\exp(-x^2) \, f(x)$ between minus- and plus-infinity using the Hermite polynomials. The method calculates the approximation for the integral by summing the product of quadrature weights and the values of function $f(x)$ at critical points of x. Here is the equation for the Gauss-Laguerre quadrature:

$$\int f(x) \, dx = \Sigma \, A_i \, f(x_i) \tag{5.7}$$

To obtain the integral of a custom function, $g(x)$, you need to write that function as $f(x) = g(x) \, \exp(x^2)$ and then use $f(x)$ in the integration process. The number of added terms (each term is the product of the weight and function value) depends on the order of the Hermite polynomial used in approximating the integral.

Here is the algorithm for the Gauss-Hermite quadrature:

Given:

- The function $f(x)$.

Algorithm:

1. Set Sum = 0
2. Initialize the array Xk to store the critical x values of the quadrature
3. Initialize the array Ak to store the weights
4. For I = 0 to 5, add Ak[I] * f(Xk[I]) to Sum
5. Return Sum as the area for e^x f(x) between zero and infinity

The Gauss-Chebyshev Quadrature

The Gauss-Chebyshev quadrature evaluates the integral of $f(x)/\sqrt{(1-x^2)}$ between –1 to 1 using the Chebyshev polynomials. The method calculates the approximation for the integral by summing the values of function f(x) at critical points of x. Here is the equation for the Gauss-Laguerre quadrature:

$$\int f(x)\ dx = (\pi/n)\ \Sigma\ f(x_i) \tag{5.8}$$

Here is the algorithm for the Gauss-Chebyshev quadrature:

Given:

- The function f(x)
- The number of points N used to evaluate the interval

Algorithm:

1. Set Sum = 0
2. For I = 1 to N, repeat the following steps:
   - 2.1. Set $X = \cos\left(\dfrac{(2*I-1)*\pi}{2N}\right)$
   - 2.2. Add X to Sum
3. Return $\dfrac{\pi}{N*Sum}$ as the area of function $\dfrac{f(x)}{\sqrt{(1-x^2)}}$

The Romberg Method

The Romberg method is an iterative method to calculate the area by successive refinements. The method calculates a column of area estimates using the trapezoidal rule. The method then uses the so-called Richardson extrapolation to refine the first set of area estimates. Here is the algorithm for the Romberg method:

Given:

- The integration limits A and B
- The tolerance factor T

- The maximum number of iterations N
- The integrated function f(x)

Algorithm:

1. Set operation flag to goOn

2. Set the number of iterations, Iter, to 0

3. Set the number of steps, M, to 1

4. Set H = B – A

5. Set fA = $\dfrac{(f(A) + f(B))}{2}$

6. Set T[0] = h * fA

7. Set Area = T[0]

8. Repeat the next steps while the operation flag is goOn:
   8.1. Increment nIter
   8.2. Halve the value of H and double the value of M
   8.3. Set c1 = T[0]
   8.4. Set J = 0 and I = M – 1
   8.5. Increment J
   8.6. Set fnArr[I – 1] = f(A + I * H)
   8.7. If I > 1, set fnArr[I – 2] = fnArr$\left[\dfrac{M}{2 - J - 1} \right]$
   8.8. Set I = I – 2
   8.9. If I >= 1, resume at step 8.5
   8.10. Set T[0] = fA
   8.11. For I = 1 to M – 1, add fnArr[I – 1] to T[0]
   8.12. Set T[0] = T[0] * H
   8.13. For I = 2 to nIter + 1, set T[I – 1] = $\dfrac{T[I – 2] + (T[I – 2] – c1)}{(4^{\wedge}(I – 1) – 1)}$
   8.14. If |T[nIter] – Area)| <= T, then set operation flag to opConverge
   8.15. If nIter >= N, then set the operation flag to opReachedMax
   8.16. Set Area = T[nIter]

9. Return Area as the integral

The Visual Basic Source Code

Let's look at the Visual Basic source code that implements the integration methods presented in this chapter. Listing 5.1 shows the source code for the INTEGRAL.BAS module file. Listing 5.2 shows the source code for the MYINTEG.BAS module file.

☞ Throughout the book, the underscore character is used to split wrapping lines of Visual Basic declarations and statements.

Listing 5.1 The source code for the INTEGRAL.BAS module file.

```
Global Const INTEGRAL_BAD_RESULT# = -1E+30
Global Const MAX_ROMBERG_TABLE% = 52


Function AltExtSimpson (numElem As Integer, yArr() As Double, _
 h As Double) As Double

 Dim sum As Double
 Dim i As Integer, isOdd As Integer

 isOdd = (numElem Mod 2) <> 0

 If numElem < 10 Then
  AltExtSimpson = INTEGRAL_BAD_RESULT
  Exit Function
 End If

 ' decrement number of element if value is even
 If Not isOdd Then numElem = numElem - 1

 ' add terms
 sum = 0
 For i = 0 To numElem
  sum = sum + yArr(i)
 Next i

 ' calculate integral
 sum = h / 48 * (48 * sum - 31 * yArr(0) + 11 * yArr(1) - _
  5 * yArr(2) + yArr(3) + yArr(numElem - 4) - _
  5 * yArr(numElem - 3) + 11 * yArr(numElem - 2) - _
  31 * yArr(numElem - 1))

 ' adjust for even number of observations
 If Not isOdd Then
  numElem = numElem + 1
  sum = sum + h / 2 * (yArr(numElem - 2) + yArr(numElem - 1))
 End If
 AltExtSimpson = sum
End Function

Function fAltExtSimpson (xFirst As Double, xLast As Double, _
 n As Integer) As Double

 Dim h As Double
 Dim x As Double
 Dim sum As Double
 Dim i As Integer, isOdd As Integer
 isOdd = (n Mod 2) <> 0

 If n < 10 Then
  fAltExtSimpson = INTEGRAL_BAD_RESULT
  Exit Function
 End If

 ' increment n if it is even
 If Not isOdd Then n = n + 1

 ' calculate increment
 h = (xLast - xFirst) / (n - 1)

 ' add terms
 x = xFirst
 sum = 0
```

Listing 5.1 (*Continued*)

```
 For i = 0 To n - 1
  sum = sum + MyFx(x)
  x = x + h
 Next i

 ' calculate integral
 sum = h / 48 * (48 * sum - 31 * MyFx(xFirst) + _
  11 * MyFx(xFirst + h) - 5 * MyFx(xFirst + 2 * h) + _
  MyFx(xFirst + 3 * h) + MyFx(xLast - 3 * h) - _
  5 * MyFx(xLast - 2 * h) + 11 * MyFx(xLast - h) - _
  31 * MyFx(xLast))

 fAltExtSimpson = sum
End Function

Function fSimpson (xFirst As Double, xLast As Double, _
 n As Integer) As Double

 Dim h As Double
 Dim x As Double
 Dim sumEven As Double
 Dim sumOdd As Double
 Dim sum As Double
 Dim i As Integer, isOdd As Integer

 isOdd = (n Mod 2) <> 0

 If n < 3 Then
  fSimpson = INTEGRAL_BAD_RESULT
  Exit Function
 End If

 ' increment n if it is even
 If Not isOdd Then n = n - 1

 ' calculate increment
 h = (xLast - xFirst) / (n - 1)

 ' add even terms
 x = xFirst + h
 sumEven = 0
 For i = 1 To n - 2 Step 2
  sumEven = sumEven + MyFx(x)
  x = x + 2 * h
 Next i

 ' add odd terms
 x = xFirst + 2 * h
 sumOdd = 0
 For i = 2 To n - 3 Step 2
  sumOdd = sumOdd + MyFx(x)
  x = x + 2 * h
 Next i

 ' calculate integral
 sum = h / 3 * (MyFx(xFirst) + 4 * sumEven + 2 * sumOdd _
  + MyFx(xLast))

 fSimpson = sum
End Function

Function GaussChebyshevQuadrature (n As Integer) As Double
```

Listing 5.1 (*Continued*)

```
Dim sum As Double
Dim x As Double
Dim pi As Double
Dim i As Integer

sum = 0
pi = 4 * Atn(1)
For i = 1 To n
 x = Cos((2 * i - 1) * pi / 2 / n)
 sum = sum + MyChebyFx(x)
Next i
GaussChebyshevQuadrature = pi / n * sum
End Function

Function GaussHermiteQuadrature () As Double

 Dim sum As Double
 Static Xk(5) As Double
 Static Ak(5) As Double
 Dim i As Integer

 Xk(0) = -2.35060497
 Xk(1) = -1.33584907
 Xk(2) = -.43607741
 Xk(3) = .43607741
 Xk(4) = 1.33584907
 Xk(5) = 2.35060497
 Ak(0) = .00453001
 Ak(1) = .15706732
 Ak(2) = .7246296
 Ak(3) = .7246296
 Ak(4) = .15706732
 Ak(5) = .00453001
 sum = 0
 For i = 0 To 5
  sum = sum + Ak(i) * MyHermiteFx(Xk(i))
 Next i
 GaussHermiteQuadrature = sum
End Function

Function GaussLaguerreQuadrature () As Double

 Dim sum As Double
 Static Xk(5) As Double
 Static Ak(5) As Double
 Dim i As Integer

 Xk(0) = .2228466
 Xk(1) = 1.1889321
 Xk(2) = 2.99273663
 Xk(3) = 5.77514357
 Xk(4) = 9.83746742
 Xk(5) = 15.98287398
 Ak(0) = .45896467
 Ak(1) = .41700083
 Ak(2) = .11337333
 Ak(3) = .0103992
 Ak(4) = .00026102
 Ak(5) = .0000009
 sum = 0
 For i = 0 To 5
  sum = sum + Ak(i) * MyLaguerreFx(Xk(i))
 Next i
```

Listing 5.1 *(Continued)*

```
 GaussLaguerreQuadrature = sum
End Function

Function GaussLegendreQuadrature (xFirst As Double, _
 xLast As Double, nSubIntervals As Integer) As Double

 Dim xA As Double, xB As Double, xJ As Double
 Dim h As Double, hDiv2 As Double
 Dim sum As Double, area As Double
 Static Xk(5) As Double
 Static Ak(5) As Double
 Dim n As Integer, i As Integer, j As Integer
 Xk(0) = -.9324695142
 Xk(1) = -.6612093865
 Xk(2) = -.2386191861
 Xk(3) = .2386191861
 Xk(4) = .6612093865
 Xk(5) = .9324695142
 Ak(0) = .1713244924
 Ak(1) = .360761573
 Ak(2) = .4679139346
 Ak(3) = .4679139346
 Ak(4) = .360761573
 Ak(5) = .1713244924
 area = 0
 n = nSubIntervals

 h = (xLast - xFirst) / n
 xA = xFirst
 For i = 1 To n
     sum = 0
     xB = xA + h
     hDiv2 = h / 2
     ' obtain area of sub-interval
     For j = 0 To 5
             xJ = xA + hDiv2 * (Xk(j) + 1)
             sum = sum + Ak(j) * MyFx(xJ)
     Next j
     area = area + hDiv2 * sum
     xA = xB
 Next i

 GaussLegendreQuadrature = area
End Function

Function Romberg (xA As Double, xB As Double, _
 tolerance As Double, MaxIter As Integer, area As Double) _
 As Integer
 Const goOn% = 1
 Const converge% = 2
 Const reachedMax% = 3

 Static T(MAX_ROMBERG_TABLE) As Double
 Static fnArr(MAX_ROMBERG_TABLE) As Double
 Dim fA As Double, h As Double, c1 As Double
 Dim i As Integer, j As Integer
 Dim nIter As Integer, nSteps As Integer
 Dim opFlag As Integer

 opFlag = goOn
 nIter = 0
 nSteps = 1
 h = xB - xA
```

Listing 5.1 *(Continued)*

```
fA = (MyFx(xA) + MyFx(xB)) / 2
T(0) = h * fA
area = T(0)

Do While opFlag = goOn
    ' divide the step size by 2
    nIter = nIter + 1
    h = h / 2
    nSteps = 2 * nSteps
    c1 = T(0)

    ' get new area estimate
    j = 0
    i = nSteps - 1
    Do
            j = j + 1
            fnArr(i - 1) = MyFx(xA + i * h)
            If i > 1 Then
              fnArr(i - 2) = fnArr(nSteps / 2 - j - 1)
     End If
            i = i - 2
    Loop While i >= 1

    T(0) = fA
    For i = 1 To nSteps - 1
            T(0) = T(0) + fnArr(i - 1)
    Next i
    T(0) = T(0) * h

    ' perform Richardson's extrapolation
    For i = 2 To (nIter + 1)
     T(i - 1) = T(i - 2) + (T(i - 2) - c1) / (4 ^ (i - 1) - 1)
    Next i

    ' verify convergence
    If Abs(T(nIter) - area) <= tolerance Then
            opFlag = converge
    ElseIf nIter >= MaxIter Then
            opFlag = reachedMax
    End If
    area = T(nIter)
 Loop
 Romberg = (opFlag = converge)
End Function

Function Simpson (numElem As Integer, yArr() As Double, _
 h As Double) As Double

 Dim sumEven As Double
 Dim sumOdd As Double
 Dim sum As Double
 Dim i As Integer, isOdd As Integer

 isOdd = (numElem Mod 2) <> 0

 If numElem < 3 Then
  Simpson = INTEGRAL_BAD_RESULT
  Exit Function
 End If

 ' decrement number of element if value is even
 If Not isOdd Then numElem = numElem - 1
```

Listing 5.1 (*Continued*)

```
' add even terms
sumEven = 0
For i = 1 To numElem - 2 Step 2
  sumEven = sumEven + yArr(i)
Next i

' add odd terms
sumOdd = 0
For i = 2 To numElem - 3 Step 2
  sumOdd = sumOdd + yArr(i)
Next i

' calculate integral
sum = h / 3 * (yArr(0) + 4 * sumEven + 2 * sumOdd + _
  yArr(numElem - 1))

' adjust for even number of observations
If Not isOdd Then
  numElem = numElem + 1
  sum = sum + h / 2 * (yArr(numElem - 2) + yArr(numElem - 1))
End If
Simpson = sum
End Function
```

Listing 5.1 declares the following Visual Basic functions:

1. The function Simpson calculates the integral using Simpson's one-third rule with an array of function values. The function returns the value of the area and has parameters that specify the number of array elements, the array of function values, and the increment in x.

2. The function fSimpson calculates the integral using Simpson's one-third rule with direct function evaluation. The function returns the value of the area and has parameters that specify the first and last value of x, and the number of subintervals.

3. The function AltExtSimpson calculates the integral using Simpson's alternate extended rule with an array of function values. The function returns the value of the area and has parameters that specify the number of array elements, the array of function values, and the increment in x.

4. The function fAltExtimpson calculates the integral using Simpson's alternate extended rule with direct function evaluation. The function returns the value of the area and has parameters that specify the first and last value of x, and the number of subintervals.

5. The function GaussLegendreQuadrature implements the Gauss-Legendre quadrature. The function returns the value of the integral and has parameters that specify the integration range and the number of subintervals.

6. The function GaussLaguerreQuadrature implements the Gauss-Laguerre quadrature. The function returns the value of the integral and has no parameters. The function relies on you to define the function MyLaguerreFx.

7. The function GaussHermiteQuadrature implements the Gauss-Hermite quadrature. The function returns the value of the integral and has no parameters. The function relies on you to define the function MyHermiteFx.

8. The function GaussChebyshevQuadrature implements the Gauss-Chebyshev quadrature. The function returns the value of the integral and has no parameters. The function relies on you to define the function MyChebyshevFx.

9. The function Romberg implements the Romberg integration. The function returns an integer value that represents the success or failure of the function. The function has parameters that define the integration range, tolerance factor, maximum number of iteration, and the pointer to the calculated area. The latter parameter returns the sought integral value.

Listing 5.2 implements the user-defined Visual Basic functions called by the function declared in Listing 5.1.

Listing 5.2 The source code for the MYINTEG.BAS module file.

```
Function MyChebyFx (X As Double)
 MyChebyFx = 1
End Function

Function Myfx (X As Double)
 Myfx = Exp(X) - 3 * X * X
End Function

Function MyHermiteFx (X As Double)
 MyHermiteFx = X ^ 2
End Function

Function MyLaguerreFx (X As Double)
 MyLaguerreFx = 1
End Function
```

The Visual Basic Test Program

Let's look at a test program that applies the interpolation functions defined in Listing 5.1. Listing 5.3 shows the source code for the program project TSINTEG.MAK. To compile the test program, you need to include the files INTEGRAL.BAS and MYINTEG.BAS in your project file.

Listing 5.3 The source code associated with the form of the program project TSINTEG.MAK.

```
Const ARRAY_SIZE% = 100

Sub ExitMnu_Click ()
 End
End Sub

Sub GaussChebyshevMnu_Click ()
 Cls
 Print "Testing the Gauss-Chebyshev method"
 Print
 Print "Area from -1 to +1 = "; GaussChebyshevQuadrature(10)

End Sub

Sub GaussHermiteMnu_Click ()
 Cls
 Print "Testing the Gauss-Hermite method"
 Print
 Print "Area from 0 to infinity = "; GaussHermiteQuadrature()
```

Listing 5.3 (*Continued*)

```
End Sub

Sub GaussianMnu_Click ()
 Dim deltaX As Double
 Dim x0 As Double
 Dim x1 As Double
 Dim x As Double
 Dim area As Double
 Dim i As Integer, n As Integer

 deltaX = .01
 x0 = 0
 x1 = 1
 x = x0

 Cls
 Print "Testing Gauss-Legendre Quadrature"
 Print
 n = 1
 Print "Area from "; x0; " to "; x1; " = "; _
  GaussLegendreQuadrature(x0, x1, n); "("; n; " subintervals)"
 n = 2
 Print "Area from "; x0; " to "; x1; " = "; _
  GaussLegendreQuadrature(x0, x1, n); "("; n; " subintervals)"
 n = 10
 Print "Area from "; x0; " to "; x1; " = "; _
  GaussLegendreQuadrature(x0, x1, n); "("; n; " subintervals)"
End Sub

Sub GaussLaguerreMnu_Click ()
 Cls
 Print "Testing the Gauss-Laguerre method"
 Print
 Print "Area from 0 to infinity = "; GaussLaguerreQuadrature()
End Sub

Sub RombergMnu_Click ()
 Dim deltaX As Double
 Dim x0 As Double
 Dim x1 As Double
 Dim x As Double
 Dim area As Double
 Dim i As Integer, n As Integer

 deltaX = .01
 x0 = 0
 x1 = 1
 x = x0

 Cls
 Print "Testing Romberg's method"
 Print
 If Romberg(x0, x1, .001, 6, area) Then
   Print "Area from "; x0; " to "; x1; " = "; area
 Else
   Print "Function Romberg failed!"
 End If
End Sub

Sub SimpsonMnu_Click ()
 Const ARR_SZE% = ARRAY_SIZE + 1
 Static yArr(ARR_SZE) As Double
 Dim deltaX As Double
```

Listing 5.3 (*Continued*)

```
Dim x0 As Double
Dim x1 As Double
Dim x As Double
Dim area As Double
Dim i As Integer, n As Integer

deltaX = .01
x0 = 0
x1 = 1
x = x0
For i = 0 To ARRAY_SIZE - 1
  yArr(i) = Myfx(x)
  x = x + deltaX
Next i
Cls
Print "Testing Simpson's rule"
Print
Print "Area from "; x0; " to "; x1; " = "; _
  Simpson(ARR_SZE, yArr(), deltaX)
Print "Area from "; x0; " to "; x1; " = "; _
  fSimpson(x0, x1, ARR_SZE)
Print
Print "Testing Simpson's Alternative extended rule"
Print
Print "Area from "; x0; " to "; x1; " = "; _
  AltExtSimpson(ARR_SZE, yArr(), deltaX)
Print "Area from "; x0; " to "; x1; " = "; _
  fAltExtSimpson(x0, x1, ARR_SZE)
End Sub
```

The project uses a form that has a simple menu system but no controls. Table 5.1 shows the menu structure and the names of the menu items. The form has the caption "Integration." The menu option Test has six selections to test the various methods for numerical integration. Each one of these menu selections clears the form and then displays the values for values integrals. Thus, you can zoom in on any method by invoking its related menu selection.

TABLE 5.1 The Menu System for the TSINTEG.MAK Project

Menu caption	Name
&Test	TesMnu
Simpson's Rules	SimpsonMnu
Gaussian Quadrature	GaussianMnu
Romberg's Method	RombergMnu
Gauss-Laguerre Method	GaussLaguerreMnu
Gauss-Hermite Method	GaussHermiteMnu
Gauss-Chebyshev Method	GaussChebyshevMnu
–	N1
&Exit	ExitMnu

The test program supplies its own data and then tests the various integration methods. Listing 5.3 shows the source code associated with the form of the program project TSINTEG.MAK. The listing declares a number of test Visual Basic functions that perform the following tasks:

1. The Simpson's Rules menu selection tests the function Simpson to calculate and display the area of $f(x) = e^x - 3x^2$ for x = 0 to 1. This test uses the array yArr as one of the arguments to functions Simpson and AltExtSimpson. The menu selection also tests the function fSimpson to calculate and display the area of $f(x) = e^x - 3x^2$ for x = 0 to 1. This test uses the Visual Basic function MyFx (defined in file MYINTEG.BAS) with the functions fSimpson and fAltExtSimpson.

2. The Gaussian Quadrature tests the function GaussLegendreQuadrature to calculate and display the area under $f(x) = e^x - 3x^2$ for x = 0 to 1. The program repeats the test with one, two, and 10 subintervals.

3. The Romberg's Method menu selection tests the function Romberg to calculate and display the area under $f(x) = e^x - 3x^2$ for x = 0 to 1. This test uses the Visual Basic function MyFx (defined in file MYINTEG.BAS). The call to function Romberg obtains the result using the variable area, which is passed as the last argument to the function call.

4. The Gauss-Laguerre Method menu selection tests the function GaussLaguerre-Quadrature to calculate and display the area under function $f(x) = 1$. The call to the function GaussLaguerreQuadrature uses the Visual Basic function MyLaguer-reFx, which is defined in file MYINTEG.BAS.

5. The Gauss-Hermite Method menu selection tests the function GaussHermite-Quadrature to calculate and display the area under function $f(x) = x^2$. The call to the function GaussHermiteQuadrature uses the Visual Basic function MyHer-miteFx, which is defined in file MYINTEG.BAS.

6. The Gauss-Chebyshev Method menu selection tests the function GaussCheby-shevQuadrature to calculate and display the area under function $f(x) = 1$. The call to GaussChebyshevQuadrature uses the Visual Basic function MyChebyFx, which is defined in file MYINTEG.BAS.

Figure 5.2 shows the output of each of these functions from the test program.

```
Testing Simpson's rule
Area from 0 to 1 = 0.718282
Area from 0 to 1 = 0.718282

Testing Simpson's Alternative extended rule
Area from 0 to 1 = 0.718282
Area from 0 to 1 = 0.718282


Testing Gauss-Legendre Quadrature
```

Figure 5.2 The output of the test program for each menu selection.

```
Area from 0 to 1 = 0.718281828449632 ( 1 subinterval)
Area from 0 to 1 = 0.718281828456705 ( 2 subintervals)
Area from 0 to 1 = 0.718281828458952 ( 10 subintervals)


Testing Romberg's method

Area from 0 to 1 = 0.718405764283874


Testing the Gauss-Laguerre method

Area from 0 to infinity = 0.999999995


Testing the Gauss-Hermite method

Area from 0 to infinity = 0.886226921199412


Testing the Gauss-Chebyshev method

Area from -1 to +1 = 3.14159265358979
```

Figure 5.2 (*Continued*)

Chapter 6

Solving Ordinary
Differential Equations

Numerical analysis provides valuable tools for solving ordinary differential equations (ODE), especially those equations that cannot be solved analytically (and there are many!). The ODE methods fall into three categories of sophistication: simple, moderate, and advanced. This chapter looks at popular methods that offer moderate solution levels. You will learn about the following methods:

- The fourth-order Runge-Kutta method
- The Runge-Kutta-Gill method
- The Runge-Kutta-Fehlberg method

The Visual Basic source code in this chapter applies these three methods for solving both single and multiple differential equations.

The Runge-Kutta Method

The third-order and fourth-order Runge-Kutta methods are among the popular methods for solving differential equations, yielding results that generally have acceptable accuracy. The method starts with an initial set of values of x and y from which subsequent values of y are calculated as a solution for the differential equation. Here are the equations involved in the fourth-order Runge-Kutta method:

$$y_{n+1} = y_n + \frac{(k_1 + 2k_2 + 2k_3 + k_4)}{6} \tag{6.1}$$

$$k_1 = h\, f(x_n, y_n) \tag{6.2}$$

$$k_2 = h\,f\!\left(x_n + \frac{h}{2},\, y_n + \frac{k_1}{2}\right) \qquad\qquad (6.3)$$

$$k_3 = h\,f\!\left(x_n + \frac{h}{2},\, y_n + \frac{k_2}{2}\right) \qquad\qquad (6.4)$$

$$k_5 = h\,(x_n + h,\, y_n + k_3) \qquad\qquad (6.5)$$

The variable h is the increment in x used to solve the differential equation. Here is the algorithm for the Runge-Kutta method:

Given:

- The initial values x and y, namely, x0 and y0
- The increment value, H
- The array Yarr, which stores the solution
- the number of array elements, N
- The function f(x, y)

Algorithm:

1. Set x = x0 and y = y0
2. Repeat the subsequent tasks for I = 0 to N – 1
3. Set k1 = H * f(x, y)
4. Set $k2 = H * f\!\left(x + \dfrac{H}{2},\, y + \dfrac{k1}{2}\right)$
5. Set $k3 = H * f\!\left(x + \dfrac{H}{2},\, y + \dfrac{k2}{2}\right)$
6. Set k4 = H * f(x + H, y + k3)
7. Set $y = y + \dfrac{(k1 + 2 * (k2 + k3) + k4)}{6}$
8. Set Yarr[I] = y
9. Add H to x

You can easily modify this algorithm to solve for an array of differential equations. To do this, use arrays for the variables H, x0, y0, x, y, k1, k2, k3, k4, and f(x, y).

The Runge-Kutta-Gill Method

The Runge-Kutta-Gill method offers better control on the round-off errors of the fourth-order Runge-Kutta method. The equations that define the Runge-Kutta-Gill method are as follows:

$$y_{n+1} = y_n + \frac{(k_1 + k_2)}{6} + \frac{c_1 k_2 + c_2 k_3}{3} \qquad (6.6)$$

$$k_1 = h\, f(x_n, y_n) \qquad (6.7)$$

$$k_2 = h\, f\left(x_n + \frac{h}{2}, y_n + \frac{k_1}{2}\right) \qquad (6.8)$$

$$k_3 = h\, f\left(x_n + \frac{h}{2}, y_n + c_3 k_1 + c_4 k_2\right) \qquad (6.9)$$

$$k_4 = h\, f(x_n + h, y_n + c_5 k_2 + c_6 k_3) \qquad (6.10)$$

where

$$c_1 = \frac{\left(1 - \frac{1}{\sqrt{2}}\right)}{3}$$

$$c_2 = \frac{\left(1 + \frac{1}{\sqrt{2}}\right)}{3}$$

$$c_3 = -0.5 + \frac{1}{\sqrt{2}}$$

$$c_4 = 1 - \frac{1}{\sqrt{2}}$$

$$c_5 = \frac{-1}{\sqrt{2}}$$

$$c_6 = 1 + \frac{1}{\sqrt{2}}$$

Here is the algorithm for the Runge-Kutta-Gill method:

Given:

- The initial values x and y, namely, x0 and y0
- The increment value, H
- The array Yarr, which stores the solution
- The number of array elements, N
- The function f(x, y)

Algorithm:

1. Set x = x0 and y = y0

2. Set C1 = $\dfrac{1}{\sqrt{2}}$, C2 = 1 + C1, C3 = 1 − C1, C4 = −0.5 + C1, and C5 = C1

3. Repeat the subsequent tasks for I = 0 to N − 1

4. Set k1 = H * f(x, y)

5. Set k2 = H * f$\left(x + \dfrac{H}{2}, y + \dfrac{k1}{2}\right)$

6. Set k3 = H * f$\left(x + \dfrac{H}{2}, y + C4 * k1 + C3 * k2\right)$

7. Set k4 = H * f(x + H, y + C5 * k2 + C2 * k3)

8. Set y = $\dfrac{y + (k1 + k4)}{6} + \dfrac{(C3 * k2 + C2 * K3)}{3}$

9. Set Yarr[I] = y

10. Add H to x

You can easily modify this algorithm to solve for an array of differential equations. To do this, use arrays for the variables H, x0, y0, x, y, k1, k2, k3, k4, and f(x, y).

The Runge-Kutta-Fehlberg Method

The Runge-Kutta-Fehlberg method offers another refinement for the Runge-Kutta method. This refinement supplies you with even better accuracy than the Runge-Kutta-Gill method. The price, of course, includes more computational effort. The Runge-Kutta-Fehlberg method uses the same notation as the Gill modification, except the Fehlberg modifications are more elaborate and use five increment factors (k1 though k5) to update the value of y.

Here is the algorithm for the Runge-Kutta-Fehlberg method:

Given:

- The initial values x and y, namely, x0 and y0
- The increment value, H
- The array Yarr, which stores the solution
- The number of array elements, N
- The function f(x, y)

Algorithm:

1. Set x = x0 and y = y0

2. Repeat the subsequent tasks for I = 0 to N − 1

3. Set k1 = H * f(x, y)

4. Set k2 = H * f$\left(x + \dfrac{H}{2}, y + \dfrac{k1}{2}\right)$

5. Set $k3 = H * f\left(x + \dfrac{3}{8} * H, y + \dfrac{3}{32} * (k1 + 3 * k2)\right)$

6. Set $k4 = H * f\left(x + \dfrac{12}{13} * H, y + \dfrac{1932}{2197} * k1 - \dfrac{7200}{2197} * k2 + \dfrac{7296}{2197} * k3\right)$

7. Set $k5 = H * f\left(x + H, y + \dfrac{439}{216} * k1 - 8 * k2 + \dfrac{3680}{513} * k3 - \dfrac{845}{4104} * k4\right)$

8. Set $y = y + \dfrac{25}{216} * k1 + \dfrac{1408}{2565} * k3 + \dfrac{2197}{4104} * k4 - \dfrac{k5}{5}$

9. Set Yarr[I] = y

10. Add H to x

You can easily modify this algorithm to solve for an array of differential equations. To do this, use arrays for the variables H, x0, y0, x, y, k1, k2, k3, k4, k5, and f(x, y).

The Visual Basic Source Code

Let's look at the Visual Basic source code that implements these methods to solve single and multiple differential equations. Listing 6.1 shows the source code for the ODE.BAS module file.

☞ Throughout the book, the underscore character is used to split wrapping lines of Visual Basic declarations and statements.

Listing 6.1 The source code for the ODE.BAS module file.

```
Sub RungeKutta4 (x0 As Double, y0 As Double, deltaX As Double, _
  yArr() As Double, numElem As Integer)

  Dim x As Double, y As Double, halfDelta As Double
  Dim k1 As Double, k2 As Double
  Dim k3 As Double, k4 As Double
  Dim i As Integer

  If numElem < 1 Then Exit Sub

  x = x0
  y = y0
  halfDelta = deltaX / 2
  For i = 0 To numElem - 1
    k1 = deltaX * MyFx(x, y)
    k2 = deltaX * MyFx(x + halfDelta, y + k1 / 2)
    k3 = deltaX * MyFx(x + halfDelta, y + k2 / 2)
    k4 = deltaX * MyFx(x + deltaX, y + k3)
    y = y + (k1 + 2 * (k2 + k3) + k4) / 6
    yArr(i) = y
    x = x + deltaX
  Next i
End Sub

Sub RungeKuttaFehlberg (x0 As Double, y0 As Double, _
  deltaX As Double, yArr() As Double, numElem As Integer)

  Dim x As Double, y As Double
  Dim k1 As Double, k2 As Double
```

Listing 6.1 *(Continued)*

```
Dim k3 As Double, k4 As Double, k5 As Double
Dim i As Integer

If numElem < 1 Then Exit Sub

x = x0
y = y0
For i = 0 To numElem - 1
  k1 = deltaX * MyFx(x, y)
  k2 = deltaX * MyFx(x + deltaX / 4, y + k1 / 4)
  k3 = deltaX * MyFx(x + 3# / 8 * deltaX, y + _
    3# / 32 * (k1 + 3 * k2))
  k4 = deltaX * MyFx(x + 12# / 13 * deltaX, y + _
    1932# / 2197 * k1 - 7200# / 2197 * k2 + _
    7296# / 2197 * k3)
  k5 = deltaX * MyFx(x + deltaX, y + 439# / 216 * k1 - _
    8 * k2 + 3680# / 513 * k3 - 845# / 4104 * k4)
  y = y + 25# / 216 * k1 + 1408# / 2565 * k3 + _
    2197# / 4104 * k4 - k5 / 5
  yArr(i) = y
  x = x + deltaX
Next i
End Sub

Sub RungeKuttaGill (x0 As Double, y0 As Double, _
 deltaX As Double, yArr() As Double, numElem As Integer)

 Dim x As Double, y As Double, halfDelta As Double
 Dim k1 As Double, k2 As Double
 Dim k3 As Double, k4 As Double
 Dim c1 As Double, c2 As Double
 Dim c3 As Double, c4 As Double
 Dim c5 As Double
 Dim i As Integer

 c1 = 1 / Sqr(2)
 c2 = 1 + c1
 c3 = 1 - c1
 c4 = -.5 + c1
 c5 = -c1

 If numElem < 1 Then Exit Sub

 x = x0
 y = y0
 halfDelta = deltaX / 2
 For i = 0 To numElem - 1
   k1 = deltaX * MyFx(x, y)
   k2 = deltaX * MyFx(x + halfDelta, y + k1 / 2)
*  k3 = deltaX * MyFx(x + halfDelta, y + c4 * k1 + c3 * k2)
   k4 = deltaX * MyFx(x + deltaX, y + c5 * k2 + c2 * k3)
   y = y + (k1 + k4) / 6 + (c3 * k2 + c2 * k3) / 3
   yArr(i) = y
   x = x + deltaX
 Next i
End Sub

Sub VectRungeKutta4 (x0 As Double, y0() As Double, _
 deltaX As Double, y1() As Double, numYVar As Integer)

 Dim x As Double, halfDelta As Double
 Dim y() As Double
```

Listing 6.1 (*Continued*)

```
Dim k1() As Double, k2() As Double
Dim k3() As Double, k4() As Double
Dim i As Integer

If numYVar < 1 Then Exit Sub
' allocate local dynamic arrays
ReDim k1(numYVar)
ReDim k2(numYVar)
ReDim k3(numYVar)
ReDim k4(numYVar)
ReDim y(numYVar)

x = x0
For i = 0 To numYVar - 1
  y(i) = y0(i)
Next i
halfDelta = deltaX / 2
For i = 0 To numYVar - 1
  k1(i) = deltaX * MyArrFx(x, y(), i)
Next i
For i = 0 To numYVar - 1
  y(i) = y(i) + k1(i) / 2
Next i
For i = 0 To numYVar - 1
  k2(i) = deltaX * MyArrFx(x + halfDelta, y(), i)
Next i
For i = 0 To numYVar - 1
  y(i) = y0(i) + k2(i) / 2
Next i
For i = 0 To numYVar - 1
  k3(i) = deltaX * MyArrFx(x + halfDelta, y(), i)
Next i
For i = 0 To numYVar - 1
  y(i) = y0(i) + k3(i)
Next i
For i = 0 To numYVar - 1
  k4(i) = deltaX * MyArrFx(x + deltaX, y(), i)
Next i
For i = 0 To numYVar - 1
  y1(i) = y0(i) + (k1(i) + 2 * (k2(i) + k3(i)) + k4(i)) / 6
Next i
End Sub

Sub VectRungeKuttaFehlberg (x0 As Double, y0() As Double, _
deltaX As Double, y1() As Double, numYVar As Integer)

Dim x As Double, halfDelta As Double
Dim y() As Double
Dim k1() As Double, k2() As Double
Dim k3() As Double, k4() As Double
Dim k5() As Double
Dim i As Integer

If numYVar < 1 Then Exit Sub

' allocate local dynamic arrays
ReDim k1(numYVar)
ReDim k2(numYVar)
ReDim k3(numYVar)
ReDim k4(numYVar)
ReDim k5(numYVar)
ReDim y(numYVar)
```

Listing 6.1 (*Continued*)

```
x = x0
For i = 0 To numYVar - 1
  y(i) = y0(i)
Next i
For i = 0 To numYVar - 1
  k1(i) = deltaX * MyArrFx(x, y(), i)
Next i
For i = 0 To numYVar - 1
  y(i) = y0(i) + k1(i) / 4
Next i
For i = 0 To numYVar - 1
  k2(i) = deltaX * MyArrFx(x + deltaX / 4, y(), i)
Next i
For i = 0 To numYVar - 1
  y(i) = y0(i) + 3# / 32 * (k1(i) + 3 * k2(i))
Next i
For i = 0 To numYVar - 1
  k3(i) = deltaX * MyArrFx(x + 3# / 8 * deltaX, y(), i)
Next i
For i = 0 To numYVar - 1
  y(i) = y0(i) + 1932# / 2197 * k1(i) - _
     7200# / 2197 * k2(i) + 7296# / 2197 * k3(i)
Next i
For i = 0 To numYVar - 1
  k4(i) = deltaX * MyArrFx(x + 12# / 13 * deltaX, y(), i)
Next i
For i = 0 To numYVar - 1
  y(i) = y0(i) + 439# / 216 * k1(i) - 8 * k2(i) + _
     3680# / 513 * k3(i) - 845# / 4104 * k4(i)
Next i
For i = 0 To numYVar - 1
  k5(i) = deltaX * MyArrFx(x + deltaX, y(), i)
Next i
For i = 0 To numYVar - 1
  y1(i) = y0(i) + 25# / 216 * k1(i) + 1408# / 2565 * k3(i) + _
     2197# / 4104 * k4(i) - k5(i) / 5
Next i
End Sub

Sub VectRungeKuttaGill (x0 As Double, y0() As Double, _
 deltaX As Double, y1() As Double, numYVar As Integer)

 Dim x As Double, halfDelta As Double
 Dim y() As Double
 Dim k1() As Double, k2() As Double
 Dim k3() As Double, k4() As Double
 Dim c1 As Double, c2 As Double
 Dim c3 As Double, c4 As Double
 Dim c5 As Double
 Dim i As Integer

 c1 = 1 / Sqr(2)
 c2 = 1 + c1
 c3 = 1 - c1
 c4 = -.5 + c1
 c5 = -c1

 If numYVar < 1 Then Exit Sub

 ' allocate local dynamic arrays
 ReDim k1(numYVar)
 ReDim k2(numYVar)
 ReDim k3(numYVar)
```

Listing 6.1 (*Continued*)

```
ReDim k4(numYVar)
ReDim y(numYVar)

x = x0
For i = 0 To numYVar - 1
  y(i) = y0(i)
Next i
halfDelta = deltaX / 2
For i = 0 To numYVar - 1
  k1(i) = deltaX * MyArrFx(x, y(), i)
Next i
For i = 0 To numYVar - 1
  y(i) = y(i) + k1(i) / 2
Next i
For i = 0 To numYVar - 1
  k2(i) = deltaX * MyArrFx(x + halfDelta, y(), i)
Next i
For i = 0 To numYVar - 1
  y(i) = y0(i) + c4 * k1(i) + c3 * k2(i)
Next i
For i = 0 To numYVar - 1
  k3(i) = deltaX * MyArrFx(x + halfDelta, y(), i)
Next i
For i = 0 To numYVar - 1
  y(i) = y0(i) + c5 * k2(i) + c2 * k3(i)
Next i
For i = 0 To numYVar - 1
  k4(i) = deltaX * MyArrFx(x + deltaX, y(), i)
Next i
For i = 0 To numYVar - 1
  y1(i) = y0(i) + (k1(i) + k4(i)) / 6 + (c3 * k2(i) + _
      c2 * k3(i)) / 3
Next i
End Sub
```

Listing 6.1 declares Visual Basic functions that implement the methods for single and multiple differential equations. The main difference between the two sets of Visual Basic functions is that the ones that solve multiple differential equations only calculate the y values for the next x value. Therefore, you need to include these Visual Basic functions in a loop to obtain a series of y values. Here are the declared Visual Basic functions:

1. The subroutine RungeKutta4 implements the fourth-order Runge-Kutta method. This subroutine has parameters that specify the starting x and y values, the increment in x, the array of y solutions, and the number of arrays elements. The subroutine RungeKutta4 uses the function MyFx, which represents the differential equation to solve. The subroutine fills the specified number of array elements with the solution of the specified differential equation.

2. The subroutine RungeKuttaGill implements the Runge-Kutta-Gill method. This subroutine has parameters that specify the starting x and y values, the increment in x, the array of y solutions, and the number of arrays elements. RungeKuttaGill uses the function MyFx, which represents the differential equation to solve, and fills the specified number of array elements with the solution of the specified differential equation.

3. The subroutine RungeKuttaFehlberg implements the Runge-Kutta-Fehlberg method. This subroutine has parameters that specify the starting x and y values, the increment in x, the array of y solutions, and the number of arrays elements. RungeKuttaFehlberg uses the function MyFx, which represents the differential equation to solve, and fills the specified number of array elements with the solution of the specified differential equation.

4. The subroutine VectRungeKutta4 implements a vector-version of the fourth-order Runge-Kutta method. This subroutine has parameters that specify the array's starting x and y values, the increment in x, the array of y solutions, and the number of differential equations. VectRungeKutta4 uses the function MyArrFx, which represents the set of differential equations to solve. The Visual Basic subroutine stores the values of y for the next x value in the parameter y1.

5. The subroutine VectRungeKuttaGill implements a vector-version of the fourth-order Runge-Kutta-Gill method. The subroutine has parameters that specify the array's starting x and y values, the increment in x, the array of y solutions, and the number of differential equations. VectRungeKuttaGill uses the function MyArrFx, which represents the set of differential equations to solve. The subroutine stores the values of y for the next x value in the parameter y1.

6. The subroutine VectRungeFehlberg implements a vector-version of the fourth-order Runge-Kutta-Fehlberg method. This subroutine has parameters that specify the array's starting x and y values, the increment in x, the array of y solutions, and the number of differential equations. VectRungeKuttaFehlberg uses the function MyArrFx, which represents the set of differential equations to solve. It stores the values of y for the next x value in the parameter y1.

Listing 6.2 shows the source code for the MYODE.BAS module file, which implements the Visual Basic functions used by the functions in Listing 6.1. The MYODE .BAS file contains the user-defined functions that represent the ordinary differential equations.

Listing 6.2 The source code for the MYODE.BAS module file.

```
Function MyArrFx (X As Double, Y() As Double, I
As Integer) As Double

  Select Case I
   Case 0
    MyArrFx = Y(1)

   Case 1
    MyArrFx = Y(0) + X

   Case Else
     MyArrFx = 1
  End Select
End Function

Function MyFx (X As Double, Y As Double) As Double
  MyFx = -X * Y ^ 2
End Function
```

The function MyArrFx in Listing 6.2 solves simultaneous ordinary differential equations. You can extend the function MyArrFx by adding more Case labels to accomodate more ODEs.

The Visual Basic Test Program

Let's look at a test program that applies the interpolation functions defined in Listing 6.1. Listing 6.3 shows the source code for the program project TSODE.MAK. To compile the test program, you need to include the files ODE.BAS and MYODE.BAS in your project file.

The project uses a form that has a simple menu system but no controls. Table 6.1 shows the menu structure and the names of the menu items. The form has the caption "Ordinary Differential Equations." The menu option Test has six selections to test the various methods for solving ODEs. Each one of these menu selections clears the form, and then displays the results of solving the ODEs. Thus, you can zoom in on any method by invoking its related menu selection.

The test program supplies its own data and then tests the various integration methods. Listing 6.3 shows the source code associated with the form of the program project TSODE.MAK.

Listing 6.3 The source code associated with the form of project TSODE.MAK.

```
Sub ExitMnu_Click ()
 End
End Sub

Sub RK2Mnu_Click ()
 Const ARRAY_SIZE% = 100
 Const NUM_SIMUL% = 2
 Dim x0 As Double
 Dim xn As Double
 Dim y0 As Double
 Dim deltaX As Double
 Static yArr0(NUM_SIMUL) As Double
 Static yArr1(NUM_SIMUL) As Double
 Dim i As Integer

 Cls
 Print "Testing the Runge-Kutta method to solve 2 equations"
 Print
 x0 = 0
 yArr0(0) = 1
 yArr0(1) = -1
 xn = 2
 deltaX = .2
 Print " X"; Tab(18); "Y0"; Tab(32); "Y1"
 Print "------"; Tab(15); "----------"; Tab(30); "----------"
 Print Format$(x0, "0.0"); Tab(15); Format$(yArr0(0), _
  "0.0000"); Tab(30); Format$(yArr0(1), "0.0000")
 Do
  VectRungeKutta4 x0, yArr0(), deltaX, yArr1(), NUM_SIMUL
  x0 = x0 + deltaX
  For i = 0 To NUM_SIMUL - 1
   yArr0(i) = yArr1(i)
  Next i
  Print Format$(x0, "0.0"); Tab(15); Format$(yArr0(0), _
```

Listing 6.3 *(Continued)*

```
   "0.0000"); Tab(30); Format$(yArr0(1), "0.0000")
 Loop While x0 < xn
End Sub

Sub RKF2Mnu_Click ()
 Const ARRAY_SIZE% = 100
 Const NUM_SIMUL% = 2
 Dim x0 As Double
 Dim xn As Double
 Dim y0 As Double
 Dim deltaX As Double
 Static yArr0(NUM_SIMUL) As Double
 Static yArr1(NUM_SIMUL) As Double
 Dim i As Integer

 Cls
 Print "Testing the Runge-Kutta-Fehlberg method to solve 2 equations"
 Print
 x0 = 0
 yArr0(0) = 1
 yArr0(1) = -1
 xn = 2
 deltaX = .2
 Print " X"; Tab(18); "Y0"; Tab(32); "Y1"
 Print "------"; Tab(15); "----------"; Tab(30); "----------"
 Print Format$(x0, "0.0"); Tab(15); Format$(yArr0(0), _
  "0.0000"); Tab(30); Format$(yArr0(1), "0.0000")
 Do
  VectRungeKuttaFehlberg x0, yArr0(), deltaX, _
             yArr1(), NUM_SIMUL
  x0 = x0 + deltaX
  For i = 0 To NUM_SIMUL - 1
   yArr0(i) = yArr1(i)
  Next i
  Print Format$(x0, "0.0"); Tab(15); Format$(yArr0(0), _
   "0.0000"); Tab(30); Format$(yArr0(1), "0.0000")
 Loop While x0 < xn
End Sub

Sub RKFMnu_Click ()
 Const ARRAY_SIZE% = 100
 Dim x0 As Double
 Dim xn As Double
 Dim y0 As Double
 Dim deltaX As Double
 Static yArr(ARRAY_SIZE) As Double

 Cls
 Print "Testing the Runge-Kutta-Fehlberg method"
 Print
 x0 = 2
 y0 = 1
 xn = 3
 deltaX = .001
 Print " X"; Tab(18); "Y"
 Print "------"; Tab(15); "----------"
 Print Format$(x0, "#.0"); Tab(15); Format$(y0, "0.0000")
 Do
  RungeKuttaFehlberg x0, y0, deltaX, yArr(), ARRAY_SIZE
  x0 = x0 + ARRAY_SIZE * deltaX
  y0 = yArr(ARRAY_SIZE - 1)
  Print Format$(x0, "#.0"); Tab(15); Format$(y0, "0.0000")
 Loop While x0 < xn
End Sub
```

Listing 6.3 *(Continued)*

```
Sub RKG2Mnu_Click ()
 Const ARRAY_SIZE% = 100
 Const NUM_SIMUL% = 2
 Dim x0 As Double
 Dim xn As Double
 Dim y0 As Double
 Dim deltaX As Double
 Static yArr0(NUM_SIMUL) As Double
 Static yArr1(NUM_SIMUL) As Double
 Dim i As Integer

 Cls
 Print "Testing the Runge-Kutta-Gill method to solve 2 equations"
 Print
 x0 = 0
 yArr0(0) = 1
 yArr0(1) = -1
 xn = 2
 deltaX = .2
 Print " X"; Tab(18); "Y0"; Tab(32); "Y1"
 Print "------"; Tab(15); "----------"; Tab(30); "----------"
 Print Format$(x0, "0.0"); Tab(15); Format$(yArr0(0), _
  "0.0000"); Tab(30); Format$(yArr0(1), "0.0000")
 Do
  VectRungeKuttaGill x0, yArr0(), deltaX, yArr1(), NUM_SIMUL
  x0 = x0 + deltaX
  For i = 0 To NUM_SIMUL - 1
   yArr0(i) = yArr1(i)
  Next i
  Print Format$(x0, "0.0"); Tab(15); Format$(yArr0(0), _
   "0.0000"); Tab(30); Format$(yArr0(1), "0.0000")
 Loop While x0 < xn
End Sub

Sub RKGMnu_Click ()
 Const ARRAY_SIZE% = 100
 Dim x0 As Double
 Dim xn As Double
 Dim y0 As Double
 Dim deltaX As Double
 Static yArr(ARRAY_SIZE) As Double

 Cls
 Print "Testing the Runge-Kutta-Gill method"
 Print
 x0 = 2
 y0 = 1
 xn = 3
 deltaX = .001
 Print " X"; Tab(18); "Y"
 Print "------"; Tab(15); "---------"
 Print Format$(x0, "#.0"); Tab(15); Format$(y0, "0.0000")
 Do
  RungeKuttaGill x0, y0, deltaX, yArr(), ARRAY_SIZE
  x0 = x0 + ARRAY_SIZE * deltaX
  y0 = yArr(ARRAY_SIZE - 1)
  Print Format$(x0, "#.0"); Tab(15); Format$(y0, "0.0000")
 Loop While x0 < xn
End Sub

Sub RKMnu_Click ()
 Const ARRAY_SIZE% = 100
 Dim x0 As Double
 Dim xn As Double
```

Listing 6.3 (*Continued*)

```
Dim y0 As Double
Dim deltaX As Double
Static yArr(ARRAY_SIZE) As Double

Cls
Print "Testing the Runge-Kutta method"
Print
x0 = 2
y0 = 1
xn = 3
deltaX = .001
Print " X"; Tab(18); "Y"
Print "------"; Tab(15); "---------"
Print Format$(x0, "#.0"); Tab(15); Format$(y0, "0.0000")
Do
  RungeKutta4 x0, y0, deltaX, yArr(), ARRAY_SIZE
  x0 = x0 + ARRAY_SIZE * deltaX
  y0 = yArr(ARRAY_SIZE - 1)
  Print Format$(x0, "#.0"); Tab(15); Format$(y0, "0.0000")
Loop While x0 < xn
End Sub
```

TABLE 6.1 The Menu System for the TSODE.MAK Project

Menu caption	Name
&Test	TesMnu
Runge-Kutta Method	RKMnu
Runge-Kutta-Gill Method	RKGMnu
Runge-Kutta-Fehlberg Method	RKFMnu
Runge-Kutta Method (Two Eqns)	RK2Mnu
Runge-Kutta-Gill Method (Two Eqns)	RKG2Mnu
Runge-Kutta-Fehlberg Method (Two Eqns)	RKF2Mnu
–	N1
&Exit	ExitMnu

The program performs the following tasks:

1. The Runge-Kutta Method menu selection tests the function RungeKutta4 to solve the differential equation $dy/dx = -xy^2$. The program supplies (2, 1) as the starting point, and uses the increment of 0.001. The test uses a do-while loop to call function RungeKutta4 while x is less than three. Thus, the solution enjoys a very good level of accuracy. The program stores the data for y in the array yArr and displays only the last value of that array in each iteration.

2. The Runge-Kutta-Gill Method menu selection tests the function RungeKuttaGill in a manner very similar to that for RungeKutta4.

3. The Runge-Kutta-Fehlberg Method menu selection tests the function RungeKutta-Fehlberg in a manner very similar to that for RungeKutta4.

4. The Runge-Kutta Method (Two Eqns) menu selection tests the function Vect-RungeKutta4 to solve the two differential equations $dy/dx = y_1$ and $dy/dx = y_0 + x$. The test supplies the initial values for x, y_0, and y_1 as 0, 1, and –1, respectively. The test uses an increment in x of 0.2, and loops while the value of x is less than two.

5. The Runge-Kutta-Gill Method (Two Eqns) menu selection tests the function Vect-RungeKuttaGill in a manner very similar to that for VectRungeKutta4.

6. The Runge-Kutta-Fehlberg Method (Two Eqns) menu selection tests the function VectRungeKuttaFehlberg in a manner very similar to that for VectRungeKutta4.

Figure 6.1 shows the output of each menu selection from the test program.

```
Testing the Runge-Kutta method

X           Y
------      ---------
2.0         1.0000
2.1         0.8299
2.2         0.7042
2.3         0.6079
2.4         0.5319
2.5         0.4706
2.6         0.4202
2.7         0.3781
2.8         0.3425
2.9         0.3120
3.0         0.2857


Testing the Runge-Kutta-Gill method

X           Y
------      ---------
2.0         1.0000
2.1         0.8299
2.2         0.7042
2.3         0.6079
2.4         0.5319
2.5         0.4706
2.6         0.4202
2.7         0.3781
2.8         0.3425
2.9         0.3120
3.0         0.2857


Testing the Runge-Kutta-Fehlberg method

X           Y
------      ---------
2.0         1.0000
2.1         0.8299
2.2         0.7042
2.3         0.6079
2.4         0.5319
2.5         0.4706
```

Figure 6.1 The output of the test program for solving ordinary differential equations.

```
2.6        0.4202
2.7        0.3781
2.8        0.3425
2.9        0.3120
3.0        0.2857


Testing the Runge-Kutta method to solve two equations

X          Y0           Y1
------     ---------    --------
0.0        1.0000       -1.0000
0.2        0.8201       -0.7987
0.4        0.6811       -0.5893
0.6        0.5855       -0.3634
0.8        0.5374       -0.1119
1.0        0.5431       0.1752
1.2        0.6106       0.5094
1.4        0.7509       0.9043
1.6        0.9774       1.3755
1.8        1.3074       1.9421
2.0        1.7621       2.6268
2.2        2.3678       3.4570


Testing the Runge-Kutta-Gill method to solve 2 equations

X          Y0           Y1
------     ---------    --------
0.0        1.0000       -1.0000
0.2        0.8201       -0.7987
0.4        0.6811       -0.5893
0.6        0.5855       -0.3634
0.8        0.5374       -0.1119
1.0        0.5431       0.1752
1.2        0.6106       0.5094
1.4        0.7509       0.9043
1.6        0.9774       1.3755
1.8        1.3074       1.9421
2.0        1.7621       2.6268
2.2        2.3678       3.4570


Testing the Runge-Kutta-Fehlberg method to solve 2 equations

X          Y0           Y1
------     ---------    --------
0.0        1.0000       -1.0000
0.2        0.8201       -0.7987
0.4        0.6811       -0.5892
0.6        0.5855       -0.3633
0.8        0.5374       -0.1119
1.0        0.5431       0.1752
1.2        0.6107       0.5095
1.4        0.7509       0.9043
1.6        0.9775       1.3756
1.8        1.3075       1.9422
2.0        1.7622       2.6269
2.2        2.3679       3.4571
```

Figure 6.1 (*Continued*)

Chapter

7

Optimization

Optimization involves finding the values of variables that yield a minimum or maximum function value. This chapter looks at methods that minimize functions with single and multiple variables. You will learn about the following methods:

- The bisection method for minimizing a single-variable function
- Newton's method for minimizing a single-variable function
- The Golden Section search method for minimizing a single-variable function
- The quadratic interpolation method for minimizing a single-variable function
- The cubic interpolation method for minimizing a single-variable function
- The flexible simplex method for minimizing a multivariable function
- Newton's method for sequential optimization of a multivariable function

The Bisection Method

The bisection method for finding a minimum works much like the bisection method that locates a root of a function. The method starts with an interval that contains the minimum, and then halves that interval to zoom in on the minimum value. Here is the algorithm for the bisection method:

Given:

- The interval [A, B], which contains the minimum value for function $f(x)$
- The first derivative of the minimized function, $f'(x)$
- The tolerance level, Tol

Algorithm:

1. Set Fa = f(A)
2. Set Fb = f(B)
3. Repeat the next steps until |A – B| > Tol:

    3.1. $C = \dfrac{(A + B)}{2}$

    3.2. Fc = f(C)

    3.3. if f'(C) * f'(A) > 0, then set A = C and Fc = Fa, else set B = C and Fc = Fb

4. Return the minimum value as C

Figure 7.1 depicts an iteration of the bisection method.

Newton's Method

You can use Newton's root-seeking method to find a minimum, maximum, and saddle point because the derivative of the targeted function is zero at these points. Here is the algorithm for Newton's method:

Given:

- The minimized function, f(x)
- The initial guess for the minimum, X
- The tolerance level, Tol
- The maximum number of iterations, N

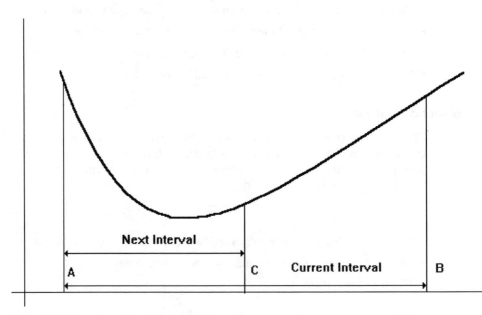

Figure 7.1 An iteration of the bisection method.

Algorithm:

1. Set Iter = 0
2. Repeat the following steps until |diff| < Tol or Iter > N:
   2.1. If |X| > 1, then set H = 0.01 * X, else set H = 0.01
   2.2. Set Fm = f(X – H)
   2.3. Set F0 = f(X)
   2.4. Set Fp = f(X + H)
   2.5. Set FirstDeriv = $\dfrac{(Fp - Fm)}{(2 * H)}$
   2.6. Set SecondDeriv = $\dfrac{(Fp - 2 * F0 + Fm)}{(H * H)}$
   2.7. Set diff = $\dfrac{FirstDeriv}{SecondDeriv}$
   2.8. Subtract diff from X
   2.9. Increment Iter
3. Return X as the minimum if Iter < N, else return an error code

This version of the algorithm approximates the first and second derivatives of the targeted function.

The Golden Section Search Method

The Golden Section search method uses the same basic principle as the bisection method in locating the minimum in an interval. The Golden Section search method uses an interval reduction factor based on the Fibonacci numbers instead of selecting the mean interval value. Here is the algorithm for the Golden Section search method:

Given:

- The interval [A, B], which contains the minimum value for function f(x)
- The tolerance level, Tol
- The maximum number of iterations, N

Algorithm:

1. Set Iter = 0
2. Set t = $\dfrac{\left(\sqrt{5} - 1\right)}{2}$
3. Set C = A + (1 – t) * (B – A)
4. Set Fc = f(C)
5. Set D = B – (1 – t) * (B – A)
6. Set Fd = f(D)

7. Repeat the next steps until |B − A| > Tol and Iter < N:
   7.1. Increment Iter++
   7.2. If Fc < Fd, then set B = D, D = C, C = A + (1 − t) * (B − A), Fd = Fc, and Fc = f(C), else set A = C, C = D, D = B − (1 − t) * (B − A), and Fc = Fc = Fd, and Fd = f(D)

8. Return $\dfrac{(A + B)}{2}$ as the minimum if Iter < N, else return an error code

Figure 7.2 depicts an iteration of the Golden Search method.

The Quadratic Interpolation Method

The quadratic interpolation method attempts to locate the minimum of a function by zooming in on the minimum instead of reducing the interval that contains the minimum. The method uses three guesses for the minimum for the interpolation. Here is the algorithm for the quadratic interpolation method:

Given:

- The points A, B, and C, which are initial guesses for the minimum of function f(x)
- The tolerances for the X values, Xtol
- The tolerance for the function, Ftol
- The maximum number of iterations, N

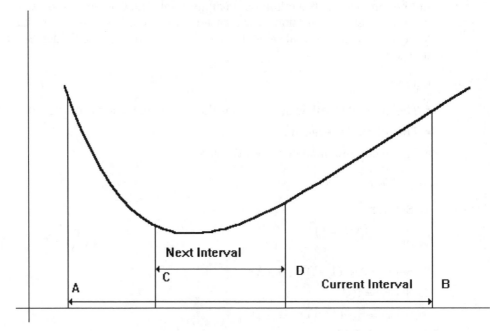

Figure 7.2 An iteration of the Golden Search method.

Algorithm:

1. Set Iter = 0
2. Set IterFlag to false
3. Set Fa = f(A)
4. Set Fb = f(B)
5. Set Fc = f(C)
6. Repeat the following steps until (IterFlag is true or $|\dfrac{(\text{LastFc} - \text{Fc})}{\text{LastFc}}| < \text{Ftol}$ and Iter > N:

    6.1. Increment Iter
    6.2. Set LastFc = Fc
    6.3. Set X =
    $$\dfrac{0.5 * ((B*B - C*C) * Fa + (C*C - A*A) * Fb + (A*A - B*B)) * Fc)}{((B - C) * Fa + (C - A) * Fb + (A - B) * Fc)}$$
    6.4. Set Fx = f(X)
    6.5. Test the following conditions:
       6.5.1. If X < C and Fx < Fc, then set B = C, C = X, Fb = Fc, and Fc = Fx
       6.5.2. Else If X > C and Fx > Fc, then set B = x and Fb = Fx
       6.5.3. Else If X < C and Fx > Fc, then set A = X and Fa = Fx
       6.5.4. Else set A = C, C = X, Fa = Fc, and Fc = Fx
    6.6. If |A − C| < Xtol or |C − B| < Xtol, then set IterFlag to true

7. Return X as the minimum if Iter <= N, else return an error code

The Cubic Interpolation Method

The cubic interpolation method attempts to locate the minimum of a function by zooming in on the minimum instead of reducing the interval that holds the minimum. The method uses two guesses for the minimum and estimates for the slope at these points to perform the interpolation. Here is the algorithm for the cubic interpolation method:

Given:

- The points A and B, which are initial guesses for the minimum of function f(x)
- The tolerances for the X values, Xtol
- The tolerance for the first derivative, Gtol
- The maximum number of iterations, N

Algorithm:

1. Set Iter = 0
2. Set Fa = f(A)
3. Set Fb = f(B)
4. Set Ga = f'(A)

5. Set Gb = f'(B)

6. Repeat the next steps until IB − AI < Xtol or Gmin < Gtol and Iter > N:

   6.1. Increment Iter

   6.2. Set $w = \dfrac{3}{(B - A) * (Fa - Fb) + Ga + Gb}$

   6.3. Set $v = \sqrt{(w * w - Ga * Gb)}$

   6.4. Set $X = \dfrac{A + (B - A) * (1 - (Gb + v - w)}{(Gb - Ga + 2 * v))}$

   6.5. Set Fx = f(X)

   6.6. Set Gx = f'(X)

   6.7. If Ga < 0 and (Gx > 0 or Fx > Fa), then set B = X, Fb = Fx, and Gb = Gx, else set A = X, Fa = Fx, and Ga = Gx

   6.8. If IGaI > IGbI, then set Gmin = Gb, else set Gmin = Ga

7. Return X as the minimum if Iter <= N, else return an error code

The Flexible Simplex Method

The flexible simplex method allows you to find the minimum for a function with multiple variables. The method obtains the minimum for a function with N variables by examining the function values at N + 1 points. This method locates the points with the best and worst values and then attempts to replace the worst-value point with a better point. This replacement process involves expanding and contracting the simplex near the worst-value point to determine a better replacement point. The iterations of the method shrinks the multidimensional simplex around the minimum point.

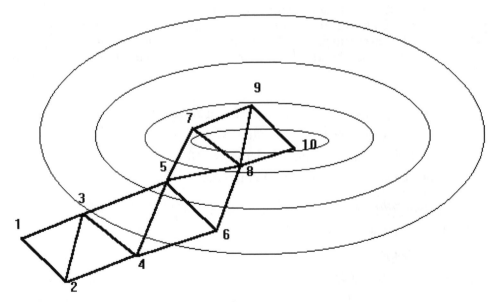

Figure 7.3 The iterations of the simplex method.

Figure 7.3 depicts the iterations of the simplex method. Here is the algorithm for the simplex method:

Given:

- N variables that evaluate function f(X)
- The matrix X, which contains N + 1 rows and N columns
- The array of function values, Y (containing N + 1 values)
- The tolerance factor, Tol
- The maximum number of iterations, M
- The reflection factor, which is a positive value that may be unity
- The expansion factor, which is greater than one (usually two)
- The contraction factor, a value between zero and one (usually 0.5)

Algorithm:

1. Set Iter = 0
2. Set P = N + 1
3. Create the dynamic arrays X1, X2, and Centroid to contain N elements
4. Calculate the function values for array Y
5. Set the convergence flag to false
6. Repeat the next steps while Iter < M and the convergence flag is false:
   6.1. Increment Iter
   6.2. Find the indices of the best and worst points
   6.3. Calculate the array of centroids for the various points, except the worst point, and store the array in Centroid
   6.4. Calculate the reflected point: for i = 0 to N − 1, set X1[i] = (1 + reflectionFactor) * Centroid[i] − reflectionFactor * X[worstI, i]
   6.5. Set Y1 = function value at point X1
   6.6. If Y1 < best Y, perform the following tasks:
      6.6.1. Calculate the expanded point: for i = 0 to N − 1, set X2[i] = (1 + expansionFactor) * X1[i] − expansionFactor * Centroid[i]
      6.6.2. Set Y2 = function value at point X2
      6.6.3. If Y2 < best Y, then replace worst point X with point X2, else replace worst point X with point X1; resume at step 7
   6.7. If Y1 >= best Y, test if Y1 is not greater than the values of array Y (except the worst Y) replace the worst X with X1 and resume at step 7; else perform the next steps:
      6.7.1. If Y1 > worst Y, replace worst X with point X1
      6.7.2. Calculate contracted point: for i = 0 to N − 1, set X2[i] = contractionFactor * X[worstI, i] + (1 − contractionFactor) * Centroid[i]
      6.7.3. Set Y2 = function value at point X2
      6.7.4. If Y2 <= worst Y, replace worst X by point X2, else for i = 0 to N − 1, set X[i] = (X[i] + best X)/2

7. Calculate array Y for current matrix X

8. Set Ymean = mean value of array Y

9. Set sum = 0

10. For i = 0 to N − 1, add $(Y[i] = Ymean)^2$ to sum

11. If $\sqrt{\dfrac{sum}{N}}$ < tolerance, set the convergence flag to true, else set the convergence flag to false

12. Move best X and Y to index 0

13. Delete the dynamic arrays X1, X2, and Centroid

Newton's Method for Sequential Optimization

It is possible to extend a single-variable optimization method, such as Newton's method, to work with multiple variables. The extended method performs repeated sequential optimization on the variables, one at a time. At the end of each cycle, the method tests for convergence.

The Visual Basic Source Code

Let's look at the Visual Basic source code that implements the optimization library. Listing 7.1 shows the source code for the OPTIM.BAS module file.

Throughout the book, the underscore character is used to split wrapping lines of Visual Basic declarations and statements.

Listing 7.1 The source code for the OPTIM.BAS module file.

```
Global Const OPTIM_BAD_RESULT# = -1E+31

Function BisectionMin (Xa As Double, Xb As Double, _
  tolerance As Double) As Double

  Dim Xc As Double
  Dim Fa As Double, Fb As Double, Fc As Double

  Fa = MyFx(Xa)
  Fb = MyFx(Xb)
  Do
   Xc = (Xa + Xb) / 2
   Fc = MyFx(Xc)
   If slope(Xc) * slope(Xa) > 0 Then
    Xa = Xc
    Fc = Fa
   Else
    Xb = Xc
    Fc = Fb
   End If
  Loop While Abs(Xb - Xa) > tolerance
  BisectionMin = Xc
End Function

Sub CalcYSimplex (X() As Double, Y() As Double, _
  numVars As Integer)
```

Listing 7.1 (*Continued*)

```
 Dim numPoints As Integer
 Dim j As Integer, i As Integer
 Dim Xarr() As Double

 numPoints = numVars + 1
 ReDim Xarr(numVars)

 ' calculate the values for y() using sumX()
 For i = 0 To numPoints - 1
  For j = 0 To numVars - 1
   Xarr(j) = X(i, j)
  Next j
  Y(i) = MySimplexFx(Xarr())
 Next i
End Sub

Function CubicIntMin (Xa As Double, Xb As Double, _
 Xtol As Double, Gtol As Double, maxIter As Integer) As Double
 Dim Fa As Double, Fb As Double, Fx As Double
 Dim Ga As Double, Gb As Double, Gx As Double
 Dim w As Double, X As Double, v As Double
 Dim h As Double, Gmin As Double
 Dim iter As Integer

 iter = 0
 Fa = MyFx(Xa)
 Fb = MyFx(Xb)
 ' calculate slope at Xa
 Ga = slope(Xa)
 ' calculate slope at Xb
 Gb = slope(Xb)
 Do
  iter = iter + 1
  w = 3 / (Xb - Xa) * (Fa - Fb) + Ga + Gb
  v = Sqr(w ^ 2 - Ga * Gb)
  X = Xa + (Xb - Xa) * (1 - (Gb + v - w) / (Gb - Ga + 2 * v))
  Fx = MyFx(X)
  ' calculate slope at x
  Gx = slope(X)
  If Ga < 0 And (Gx > 0 Or Fx > Fa) Then
   Xb = X
   Fb = Fx
   Gb = Gx
  Else
   Xa = X
   Fa = Fx
   Ga = Gx
  End If
  If Abs(Ga) > Abs(Gb) Then
   Gmin = Abs(Gb)
  Else
   Gmin = Abs(Ga)
  End If
 Loop While Not (Abs(Xb - Xa) < Xtol Or Gmin < Gtol Or _
      iter > maxIter)
 If iter <= maxIter Then
  CubicIntMin = X
 Else
  CubicIntMin = OPTIM_BAD_RESULT
 End If
End Function
```

Listing 7.1 *(Continued)*

```
Function ExNewtonMin (X() As Double, tolerance As Double, _
 diff As Double, maxIter As Integer, index As Integer) _
 As Integer

 Dim h As Double
 Dim Fm As Double, F0 As Double
 Dim Fp As Double
 Dim firstDeriv As Double, secondDeriv As Double
 Dim iter As Integer
 Dim xOld As Double
 Dim xx As Double

 iter = 0
 xOld = X(index)
 xx = xOld
 Do
  ' calculate increment
  If Abs(xx) > 1 Then h = .01 * xx Else h = .01
  ' calculate function values at x-h, X, and x+h
  X(index) = xx - h
  Fm = MySimplexFx(X())
  X(index) = xx + h
  Fp = MySimplexFx(X())
  X(index) = xx
  F0 = MySimplexFx(X())
  ' calculate the first derivative
  firstDeriv = (Fp - Fm) / 2 / h
  ' calculate the second derivative
  secondDeriv = (Fp - 2 * F0 + Fm) / h ^ 2
  ' calculate the guess refinement
  diff = firstDeriv / secondDeriv
  ' refine the guess
  xx = xx - diff
  X(index) = xx
  iter = iter + 1 ' the increment iteration counter
 Loop While Abs(diff) > tolerance And iter < maxIter

 If iter <= maxIter Then
  ExNewtonMin = True
 Else
  X(index) = xOld
  ExNewtonMin = False
 End If
End Function

Function GoldenSearchMin (Xa As Double, Xb As Double, _
 tolerance As Double, maxIter As Integer) As Double

 Dim Xc As Double, Xd As Double
 Dim Fc As Double, Fd As Double
 Dim oneMinusTau As Double
 Dim iter As Integer

 iter = 0
 oneMinusTau = 1 - (Sqr(5) - 1) / 2
 Xc = Xa + oneMinusTau * (Xb - Xa)
 Fc = MyFx(Xc)
 Xd = Xb - oneMinusTau * (Xb - Xa)
 Fd = MyFx(Xd)
 Do
  iter = iter + 1
  If Fc < Fd Then
    Xb = Xd
```

Listing 7.1 (*Continued*)

```
      Xd = Xc
      Xc = Xa + oneMinusTau * (Xb - Xa)
      Fd = Fc
      Fc = MyFx(Xc)
   Else
      Xa = Xc
      Xc = Xd
      Xd = Xb - oneMinusTau * (Xb - Xa)
      Fc = Fd
      Fd = MyFx(Xd)
   End If
 Loop While Abs(Xb - Xa) > tolerance And iter < maxIter
 If iter <= maxIter Then
  GoldenSearchMin = Xc
 Else
  GoldenSearchMin = OPTIM_BAD_RESULT
 End If
End Function

Function NewtonMin (X As Double, tolerance As Double, _
 maxIter As Integer) As Double

 Dim h As Double, diff As Double
 Dim Fm As Double, F0 As Double, Fp As Double
 Dim firstDeriv As Double, secondDeriv As Double
 Dim iter As Integer

 iter = 0
 Do
  ' caluclate increment
  If Abs(X) > 1 Then h = .01 * X Else h = .01
  ' calculate function values at X-h, X, and X+h
  Fm = MyFx(X - h)
  F0 = MyFx(X)
  Fp = MyFx(X + h)
  ' calculate the first derivative
  firstDeriv = (Fp - Fm) / 2 / h
  ' calculate the second derivative
  secondDeriv = (Fp - 2 * F0 + Fm) / h ^ 2
  ' calculate the guess refinement
  diff = firstDeriv / secondDeriv
  ' refine the guess
  X = X - diff
  iter = iter + 1 ' the increment iteration counter
 Loop While Abs(diff) > tolerance And iter < maxIter
 If iter <= maxIter Then
  NewtonMin = X
 Else
  NewtonMin = OPTIM_BAD_RESULT
 End If
End Function

Function NewtonMultiMin (X() As Double, numVars As Integer, _
 tolerance As Double, maxIter As Integer) As Integer

 Dim i As Integer, j As Integer
 Dim k As Integer
 Dim diff As Double
 Dim ok As Integer

 Do
  ok = False
  For i = 0 To numVars - 1
```

Listing 7.1 (*Continued*)

```
    If ExNewtonMin(X(), tolerance, diff, maxIter, i) Then
      If diff > tolerance Then
       ok = True
      End If
     Else
      NewtonMultiMin = False
      Exit Function
     End If
   Next i
 Loop While ok
 NewtonMultiMin = True
End Function

Function QuadIntMin (Xa As Double, Xb As Double, Xc As Double, _
  Xtol As Double, Ftol As Double, maxIter As Integer) As Double

 Dim Fa As Double, Fb As Double, Fc As Double
 Dim lastFc As Double, X As Double, Fx As Double
 Dim iter As Integer
 Dim ok As Integer

 iter = 0
 ok = False
 Fa = MyFx(Xa)
 Fb = MyFx(Xb)
 Fc = MyFx(Xc)

 Do
  iter = iter + 1
  lastFc = Fc
  X = .5 * ((Xb ^ 2 - Xc ^ 2) * Fa + (Xc ^ 2 - Xa ^ 2) * Fb + _
     (Xa ^ 2 - Xb ^ 2) * Fc) / ((Xb - Xc) * Fa + _
     (Xc - Xa) * Fb + (Xa - Xb) * Fc)
  Fx = MyFx(X)
  If X < Xc And Fx < Fc Then
   Xb = Xc
   Xc = X
   Fb = Fc
   Fc = Fx
  ElseIf X > Xc And Fx > Fc Then
   Xb = X
   Fb = Fx
  ElseIf X < Xc And Fx > Fc Then
   Xa = X
   Fa = Fx
  Else
   Xa = Xc
   Xc = X
   Fa = Fc
   Fc = Fx
  End If
  If Abs(Xa - Xc) < Xtol Or Abs(Xc - Xb) < Xtol Then
   ok = True
  End If
 Loop While Not (ok Or Abs((lastFc - Fc) / lastFc) < Ftol _
        And iter > maxIter)
 If iter <= maxIter Then
  QuadIntMin = X
 Else
  QuadIntMin = OPTIM_BAD_RESULT
 End If
End Function
```

Listing 7.1 (*Continued*)

```
Sub Simplex (X() As Double, Y() As Double, numVars As Integer, _
 tolerance As Double, reflectFact As Double, _
 expandFact As Double, contractFact As Double, _
 maxIter As Integer)

 Const EPS# = .00000001
 Const defReflectFact# = 1#
 Const defExpandFact# = 2#
 Const defContractFact# = .5

 Dim numIter As Integer
 Dim numPoints As Integer
 Dim j As Integer, i As Integer
 Dim worstI As Integer, bestI As Integer
 Dim goOn As Integer
 Dim flag As Integer
 Dim y1 As Double, y2 As Double, x0 As Double
 Dim yMean As Double, sum As Double
 Dim X1() As Double
 Dim X2() As Double
 Dim Centroid() As Double

 numIter = 0
 numPoints = numVars + 1
 goOn = True

 ' allocate dynamic arrays
 ReDim X1(numVars)
 ReDim X2(numVars)
 ReDim Centroid(numVars)

 ' check Simplex modification factors
 If reflectFact < EPS Then
  reflectFact = defReflectFact
 End If
 If expandFact < EPS Then
  expandFact = defExpandFact
 End If
 If contractFact < EPS Then
  contractFact = defContractFact
 End If
 ' calculate the values for y() using X1()
 For i = 0 To numPoints - 1
  For j = 0 To numVars - 1
   X1(j) = X(i, j)
  Next j
  Y(i) = MySimplexFx(X1())
 Next i

 Do While numIter < maxIter And goOn
  numIter = numIter + 1
  ' find worst and best point
  worstI = 0
  bestI = 0
  For i = 1 To numPoints - 1
   If Y(i) < Y(bestI) Then
    bestI = i
   ElseIf Y(i) > Y(worstI) Then
    worstI = i
   End If
  Next i
  ' calculate centroid (exclude worst point)
```

Listing 7.1 (*Continued*)

```
For i = 0 To numVars - 1
 Centroid(i) = 0
 For j = 0 To numPoints - 1
  If j <> worstI Then
   Centroid(i) = Centroid(i) + X(j, i)
  End If
 Next j
 Centroid(i) = Centroid(i) / numVars
Next i
' calculate reflected point
For i = 0 To numVars - 1
 X1(i) = (1 + reflectFact) * Centroid(i) - _
     reflectFact * X(worstI, i)
Next i
y1 = MySimplexFx(X1())

If y1 < Y(bestI) Then
 ' calculate expanded point
 For i = 0 To numVars - 1
  X2(i) = (1 + expandFact) * X1(i) - _
      expandFact * Centroid(i)
 Next i
 y2 = MySimplexFx(X2())
 If y2 < Y(bestI) Then
  ' replace worst point by X2
  For i = 0 To numVars - 1
   X(worstI, i) = X2(i)
  Next i
 Else
  ' replace worst point by X1
  For i = 0 To numVars - 1
   X(worstI, i) = X1(i)
  Next i
 End If
Else
 flag = True
 For i = 0 To numPoints - 1
  If i <> worstI And y1 <= Y(i) Then
   flag = False
   Exit For
  End If
 Next i
 If flag Then
  If y1 < Y(worstI) Then
   ' replace worst point by X1
   For i = 0 To numVars - 1
    X(worstI, i) = X1(i)
   Next i
   Y(worstI) = y1
  End If
  ' calculate contracted point
  For i = 0 To numVars - 1
   X2(i) = contractFact * X(worstI, i) + _
       (1 - contractFact) * Centroid(i)
  Next i
  y2 = MySimplexFx(X2())
  If y2 > Y(worstI) Then
   ' store best x in X1
   For i = 0 To numVars - 1
    X2(i) = X(bestI, i)
   Next i
   For j = 0 To numPoints - 1
    For i = 0 To numVars - 1
```

Listing 7.1 (*Continued*)

```
      X(j, i) = .5 * (X2(i) + X(j, i))
    Next i
   Next j
  Else
   ' replace worst point by X2
   For i = 0 To numVars - 1
    X(worstI, i) = X2(i)
   Next i
  End If
 Else
  ' replace worst point by X1
  For i = 0 To numVars - 1
   X(worstI, i) = X1(i)
  Next i
 End If
End If
' calculate the values for y() using X1
For i = 0 To numPoints - 1
 For j = 0 To numVars - 1
  X1(j) = X(i, j)
 Next j
 Y(i) = MySimplexFx(X1())
Next i
' calculate mean y
yMean = 0
For i = 0 To numPoints - 1
 yMean = yMean + Y(i)
Next i
yMean = yMean / numPoints
' calculate deviation from mean y
sum = 0
For i = 0 To numPoints - 1
 sum = sum + (Y(i) - yMean) ^ 2
Next i
' test convergence
If Sqr(sum / numPoints) > tolerance Then
 goOn = True
Else
 goOn = False
End If
Loop

' find the best point
bestI = 0
For i = 1 To numPoints - 1
 If Y(i) < Y(bestI) Then bestI = i
Next i
If bestI <> 0 Then
 ' move best point to index 0
 For i = 0 To numVars - 1
  x0 = X(0, i)
  X(0, i) = X(bestI, i)
  X(bestI, i) = x0
 Next i
 y1 = Y(0)
 Y(0) = Y(bestI)
 Y(bestI) = y1
End If
End Sub

Private Function slope (X As Double) As Double
 Dim h As Double
 If Abs(X) > 1 Then
```

Listing 7.1 (*Continued*)

```
 h = .01 * X
Else
 h = .01
End If
 slope = (MyFx(X + h) - MyFx(X - h)) / 2 / h
End Function
```

The file in Listing 7.1 declares the following functions:

1. The function BisectionMin applies the bisection method to return the minimum value of a function. This function has parameters that specify the interval containing the minimum, and the tolerance. BisectionMin minimizes the function MyFx, which is defined in module MYOPTIM.BAS.

2. The function NewtonMin uses Newton's method to return the minimum value of a function. The function has parameters that specify the initial guess for the minimum, the tolerance, and the maximum number of iterations. NewtonMin minimizes the function MyFx, which is defined in module MYOPTIM.BAS.

3. The function GoldenSearchMin applies the Golden Section search method to return the minimum value of a function. The function has parameters that specify the interval containing the minimum, the tolerance, and the maximum number of iterations. GoldenSearchMin minimizes the function MyFx which is defined in module MYOPTIM.BAS.

4. The function QuadIntMin uses the quadratic interpolation method to yield the minimum value of a function. The function has parameters that specify the initial three guesses for the minimum, the tolerance for the independent variable, the tolerance for the function values, and the maximum number of iterations. QuadIntMin minimizes the function MyFx, which is defined in module MYOPTIM.BAS.

5. The function CubeIntMin uses the cubic interpolation method to return the minimum value of a function. The function has parameters that specify the initial two guesses for the minimum, the tolerance for the independent variable, the tolerance for the slope values, and the maximum number of iterations. CubeIntMin minimizes the function MyFx, which is defined in module MYOPTIM.BAS.

6. The function CalcYSimplex calculates the values of the function for a given data matrix. The function has parameters that specify the independent variable's data matrix, the array of calculated function values, and the number of variables. CalcYSimplex uses the function MySimplexFx defined in module MYOPTIM.BAS.

7. The subroutine Simplex applies the Simplex method to locate the minimum value for a multivariable function. The subroutine has parameters that specify the independent variables' data matrix, the array of calculated function values, the number of variables, the tolerance, the reflection factor, the expandion factor, and the contraction factor. The subroutine Simplex minimizes the function MySimplexFx defined in module MYOPTIM.BAS. If the argument for the reflection factor, the expansion factor, or the contraction factor is zero, the subroutine assigns a default value for that factor.

8. The function NewtonMultiMin sequentially minimizes a multivariable function. The function has parameters that pass the initial guess for the minimum point, the number of variables, the tolerance level, and the maximum number of iterations. NewtonMultiMin minimizes the function MySimplexFx defined in module MYOPTIM.BAS. The function returns TRUE if it find a minimum, otherwise the function yields FALSE. The parameter x yields the point for the sought minimum function value.

The OPTIM.BAS module file contains the Visual Basic function getYTest and ExNewtonMin, which are local to the library. The function getYTest calculates a new point for the function Simplex. The function ExNewtonMin is called by function NewtonMultiMin to optimize a single variable.

Listing 7.2 shows the source code for the MYOPTIM.BAS module file. The file contains the functions MyFx and MySimplexFx, which are used by the functions in Listing 7.1.

Listing 7.2 The source code for the MYOPTIM.BAS file.

```
Function MyFx (x As Double)
  MyFx = (x - 3) ^ 2 + 1
End Function

Function MySimplexFx (x() As Double)
  MySimplexFx = (x(0) - .5) ^ 2 + (x(1) - .15) ^ 2 + 1
End Function
```

The Visual Basic Test Program

Let's look at a program that tests the Visual Basic functions in the optimization library. Listing 7.3 contains the source code for the form in the project the TSOP-TIM.MAK. To compile the test program, you need to include the files OPTIM.BAS and MYOPTIM.BAS in your project file.

Listing 7.3 The source code for the form associated with the TSOPTIM.MAK program project.

```
Sub BisectionMnu_Click ()
Dim Xa As Double, Xb As Double
Dim X As Double, Xtol As Double

 Cls
 Print "Testing the Bisection method"
 Print
 Xa = 1
 Xb = 4
 Xtol = .00001
 Print "Initial interval is ("; Xa; ","; Xb; ")"
 X = BisectionMin(Xa, Xb, Xtol)
 Print "Optimum at "; Format$(X, "0.00")
End Sub

Sub CubicMnu_Click ()
 Dim Xa As Double, Xb As Double
 Dim X As Double, Xtol As Double
 Dim Gtol As Double
 Dim maxIter As Integer
```

Listing 7.3 (*Continued*)

```
 Cls
 Print "Testing the Cubic Interpolation method"
 Print
 Xa = 1
 Xb = 5
 Xtol = .00000001
 Gtol = .01
 maxIter = 30
 Print "Initial guesses are "; Xa; " And "; Xb
 X = CubicIntMin(Xa, Xb, Xtol, Gtol, maxIter)
 Print "Optimum at "; Format$(X, "0.00")
End Sub

Sub ExitMnu_Click ()
 End
End Sub

Sub Form_Load ()
 AutoRedraw = True
End Sub

Sub GoldenSearchMnu_Click ()
 Dim Xa As Double, Xb As Double
 Dim X As Double, Xtol As Double
 Dim maxIter As Integer
 Cls
 Print "Testing the Golden Search method"
 Print
 Xa = 1
 Xb = 5
 Xtol = .00001
 maxIter = 30
 Print "Initial interval is ("; Xa; ","; Xb; ")"
 X = GoldenSearchMin(Xa, Xb, Xtol, maxIter)
 Print "Optimum at "; Format$(X, "0.00")
End Sub

Sub NewtonMnu_Click ()
 Dim X As Double
 Dim Xtol As Double
 Dim maxIter As Integer

 Cls
 Print "Testing Newton's method"
 Print
 X = 5
 Xtol = .00000001
 maxIter = 30
 Print "Initial guess is "; X
 X = NewtonMin(X, Xtol, maxIter)
 Print "Optimum at "; Format$(X, "0.00")

End Sub

Sub NewtonMultiVarMnu_Click ()
 Dim Xarr() As Double
 Dim Xtol As Double
 Dim maxIter As Integer
 Dim numVars As Integer

 ReDim Xarr(3)
 Xarr(0) = 0
 Xarr(1) = 0
```

Listing 7.3 *(Continued)*

```
Xtol = .00001
maxIter = 1000
numVars = 2

Cls
Print "Testing Newton's Extended Method"
Print
Print "Initial guesses for coordinates:"
For i = 0 To numVars - 1
  Print "X("; i + 1; ") = "; Xarr(i)
Next i
If NewtonMultiMin(Xarr(), numVars, Xtol, maxIter) Then
  Print "Minimum at:"
  For i = 0 To numVars - 1
    Print "X("; i + 1; ") = "; Format$(Xarr(i), "0.00")
  Next i
Else
  Print "Newton's method failed!"
End If
End Sub

Sub QuadraticMnu_Click ()
 Dim Xa As Double, Xb As Double
 Dim Xc As Double, Xtol As Double
 Dim X As Double, Ftol As Double
 Dim maxIter As Integer

 Cls
 Print "Testing the Quadratic Interpolation method"
 Print
 Xa = 1
 Xb = 2
 Xc = 5
 Xtol = .00000001
 Ftol = .0000001
 maxIter = 30
 Print "Initial guesses are "; Xa; ","; Xb; ", and "; Xc
 X = QuadIntMin(Xa, Xb, Xc, Xtol, Ftol, maxIter)
 Print "Optimum at "; Format$(X, "0.00")
End Sub

Sub SimplexMnu_Click ()
 Dim i As Integer, j As Integer
 Dim numVars As Integer
 Dim numPoints As Integer
 Dim mat() As Double
 Dim Y() As Double

 numVars = 2
 numPoints = numVars + 1
 ' allocate dynamic arrays
 ReDim mat(numPoints, numVars)
 ReDim Y(numPoints)

 mat(0, 0) = 1.5
 mat(0, 1) = 2#

 mat(1, 0) = -1.5
 mat(1, 1) = -2#

 mat(2, 0) = 3#
 mat(2, 1) = 1#
 WindowState = 2 ' maximize window
```

Listing 7.3 (*Continued*)

```
Cls
Print "Testing Simplex Method"
Print
Print "Initial guesses for coordinates:"
CalcYSimplex mat(), Y(), numVars
For j = 0 To numVars
 Print "Y("; j; ") = "; Y(j)
 For i = 0 To numVars - 1
  Print "X("; j; ","; i; ") = "; mat(j, i)
 Next i
Next j

Simplex mat(), Y(), numVars, .0000001, 0#, 0#, 0#, 1000
Print "Final coordinates:"
For j = 0 To numVars
 Print "Y("; j; ") = "; Format$(Y(j), "0.000")
 For i = 0 To numVars - 1
  Print "X("; j; ","; i; ") = "; Format$(mat(j, i), "0.000")
 Next i
Next j
End Sub
```

The project uses a form that has a simple menu system but no controls. Table 7.1 shows the menu structure and the names of the menu items. The form has the caption "Optimization." The menu option Test has seven selections to test the various methods for minimizing functions. Each one of these menu selections clears the form and then displays the values that minimize the tested functions. Thus, you can zoom in on any method by invoking its related menu selection.

TABLE 7.1 The Menu System for the TSOPTIM.MAK Project

Menu caption	Name
&Test	TesMnu
Bisection Method	BisectionMnu
Newton's Method	NewtonMnu
Golden Search Method	GoldenSearchMnu
Quadratic Interpolation Method	QuadraticMnu
Cubic Interpolation Method	CubicMnu
Simplex Method	SimplexMnu
Newton Multivariable Method	NewtonMultiVarMnu
–	N1
&Exit	ExitMnu

The test program supplies its own data and then tests the various optimization methods. The program locates the minimum of the following functions:

$$f(x) = (x - 3)^2 + 1$$

$$f_2(x, y) = (x - 0.5)^2 + (y - 0.15)^2 + 1$$

$$f_3(x, y, z) = (x - 0.5)^2 + (y - 0.15)^2 + (z - 1)^2 + 1$$

The mathematical functions $f(x)$, $f_2(x, y)$, and $f_3(x, y, z)$ are implemented using the Visual Basic functions f, f2, and f3, respectively.

The program performs the following tests:

1. The Bisection Method menu selection tests the function BisectionMin to get the minimum of function $f(x)$. The program supplies the interval [1, 4] and specifies a tolerance level of 1.0^{-5}.

2. The Newton's Method menu selection tests the function NewtonMin to get the minimum of function $f(x)$. The program supplies the initial guess of 4, a maximum of 30 iterations, and specifies a tolerance level of 1.0^{-5}.

3. The Golden Search Method menu selection tests the function GoldenSearchMin to get the minimum of function $f(x)$. The program supplies the interval [1, 5], a maximum of 30 iterations, and specifies a tolerance level of 1.0^{-8}.

4. The Quadratic Interpolation Method menu selection tests the function QuadInt-Min to get minimum of function $f(x)$. The program supplies the guesses 1, 2, and 5, the X tolerance level of 1.0^{-8}, the function tolerance level of 1.0^{-7}, and a maximum of 30 iterations.

5. The Cubic Interpolation Method menu selection tests the function QuadIntMin to get minimum of function $f(x)$. The program supplies the guesses 1 and 5, the X tolerance level of 1.0^{-8}, the function tolerance level of 1.0^{-7}, and a maximum of 30 iterations.

6. The Simplex Method menu selection tests the subroutine Simplex to get the minimum of function $f_2(x, y)$. The program supplies the data matrix, displays the initial values of the data matrix and the function values, solves for the minimum, and then displays the final values of the data matrix and the minimized function values.

7. The Newton Multivariable Method menu selection tests the function NewtonMul-tiMin to get the minimum of function $f_3(x, y, z)$. The program supplies the array of initial guesses, the tolerance level, the maximum number of iterations, and the number of variables. The program displays the initial guess, calls function New-tonMultiMin, and then displays the result.

Figure 7.4 shows the output of the test program for each menu selection.

```
Testing the Bisection method

Initial interval is ( 1 , 4 )
Optimum at 3.00


Testing Newton's method

Initial guess is 5
Optimum at 3.00


Testing the Golden Search method

Initial interval is ( 1 , 5 )
Optimum at 3.00


Testing the quadratic interpolation method

Initial guesses are 1 , 2 , and 5
Optimum at 3.00


Testing the cubic interpolation method

Initial guesses are 1 and 5
Optimum at 3.00


Testing the simplex method

Initial guesses for coordinates:
Y( 0 ) = 5.4225
X( 0 , 0 ) = 1.5
X( 0 , 1 ) = 2
Y( 1 ) = 9.6225
X( 1 , 0 ) = -1.5
X( 1 , 1 ) = -2
Y( 2 ) = 7.9725
X( 2 , 0 ) = 3
X( 2 , 1 ) = 1
Final coordinates:
Y( 0 ) = 1.000
X( 0 , 0 ) = 0.500
X( 0 , 1 ) = 0.150
Y( 1 ) = 1.000
X( 1 , 0 ) = 0.500
X( 1 , 1 ) = 0.151
Y( 2 ) = 1.000
X( 2 , 0 ) = 0.500
X( 2 , 1 ) = 0.150


Testing Newton's Extended Method

Initial guesses for coordinates:
X( 1 ) = 0
X( 2 ) = 0
Minimum at:
X( 1 ) = 0.50
X( 2 ) = 0.15
```

Figure 7.4 The output of the Visual Basic test program for optimization methods.

Basic Statistics

This chapter and the remaining three discuss popular statistical algorithms. This chapter discusses Visual Basic functions that perform basic statistics, which include the following:

- The mean and standard deviation statistics
- The confidence intervals for the mean and standard deviation
- The first four moments
- Statistical testing for the mean

The Mean and Standard Deviation

The mean and standard deviation statistics represent the simplest and perhaps the most common statistics about data. Here are the equations for calculating the mean and standard deviation:

$$\mu = \frac{\Sigma x}{n} \tag{8.1}$$

$$\sigma = \sqrt{\frac{(\Sigma x^2 - (\Sigma x)^2/n)}{(n-1)}} \tag{8.2}$$

where n is the number of observations and x is the observed variable. The algorithm for obtaining the mean and standard deviation is as follows:

Given:

- The array Xarr, which contains N observations

Algorithm:

1. Set SumX = 0 and SumX2 = 0
2. For I = 0 to N − 1 repeat the next steps:
   2.1. Add Xarr[I] to SumX
   2.2. Add Xarr[I] squared to SumX2
3. Set Mean = $\dfrac{\text{SumX}}{\text{N}}$
4. Set Sdev = $\sqrt{\dfrac{(\text{SumX2} - \text{SumX} * \text{SumX/N})}{(\text{N} - 1)}}$
5. Return Mean and Sdev as the mean and standard deviation values

The Confidence Intervals

The mean and standard deviations calculated for a sample data are estimates of the general population statistics. You can calculate the confidence interval for the range of the mean value using the following equation:

$$m \pm \frac{s\, t_{n-1;\alpha/2}}{\sqrt{n}} \tag{8.3}$$

where m is the calculated mean for the sample, s is the calculated standard deviation for the sample, t is the Student-t probability distribution function for n–1 degrees of freedom and 1–α probability, and n is the number of observations in the data sample.

Regarding the confidence interval of the standard deviation, the following equations specify the lower and upper limits:

$$\text{Lower limit} = \frac{(n-1)\, s2}{\chi^2_{n-1:\alpha/2}} \tag{8.4}$$

$$\text{Upper limit} = \frac{(n-1)\, s^2}{\chi^2_{n-1:1-\alpha/2}} \tag{8.5}$$

where χ^2 is the Chi-square probability distribution function.

The First Four Moments

Basic statistical analysis provides the first four moments along with the moment coefficients of skewness and kurtosis to offer a bit more insight about the distribution of a data sample. The first moment is the mean value; the second moment is the variance. The coefficient of skewness measures the lack of symmetry in a distribution. The coefficient of kurtosis is the relative peakness or flatness of a distribution.

Here are the equations that calculate the statistics in this section:

$$m_1 = \overline{x} = \frac{1}{n}\sum_{i=1}^{n} x_1 \tag{8.6}$$

$$m_2 = \frac{1}{n} \Sigma x_i^2 - \bar{x}^2 \tag{8.7}$$

$$m_3 = \frac{1}{n} \Sigma_i^3 - \frac{3}{n} \bar{x} \Sigma x_i^2 + 2\bar{x}^3 \tag{8.8}$$

$$m_4 = \frac{1}{n} \Sigma x_i^4 - \frac{4}{n} \bar{x} \Sigma x_i^3 + \frac{6}{n} \bar{x}^2 \Sigma x_i^2 - 3\bar{x}^4 \tag{8.9}$$

$$\gamma_1 = \frac{m_3}{m_2^{3/2}} \tag{8.10}$$

$$\gamma_2 = \frac{m_4}{m_2^2} \tag{8.11}$$

where γ_1 is the moment coefficient of skewness and γ_2 is the moment coefficient of kurtosis.

The algorithm for calculating the first four moments and their associated coefficients is as follows:

Given:

- The array Xarr, which contains N observations

Algorithm:

1. Set SumX = 0, SumX2 = 0, SumX3 = 0, and SumX4 = 0
2. For I = 0 to N − 1, repeat the next steps:
   2.1. Add Xarr[I] to SumX
   2.2. Add Xarr[I] squared to SumX2
   2.3. Add Xarr[I] cubed to SumX3
   2.4. Add Xarr[I] raised to the fourth power to SumX4

3. Set $M1 = \dfrac{SumX}{N}$

4. Set $M2 = \dfrac{SumX2}{(N - M1^2)}$

5. Set $M3 = \dfrac{SumX3}{N} - \left(\dfrac{3}{N} * M1 * SumX2\right) + 2 * M1^3$

6. Set $M4 = \dfrac{SumX4}{N} - \left(\dfrac{4}{N} * M1 * SumX3\right) + \left(\dfrac{6}{N} * M1^2 * SumX2 - 3 * M1^4\right)$

7. Set $Gamma1 = \dfrac{M3}{M2^{3/2}}$

8. Set $Gamma2 = \dfrac{M4}{M2^2}$

9. Return M1, M2, M3, M4, Gamma1, and Gamma2 as the sought statistics

Testing Sample Means

Statisticians offer many tests that compare the estimated values of the means and standard deviations of samples. This section looks at the test that compares two means of two data samples. The test is based on calculating the Student-t value and comparing it with a tabulated value. If the calculated value exceeds the tabulated value, then you do not accept the hypothesis that the means are equal. By contrast, if the calculated value does not exceed the tabulated value, then you do not reject the hypothesis that the means are equal. Here are the equations that perform the test:

$$t = \frac{(m_x - m_y - d)}{\sqrt{(T_1\, T_2)}} \tag{8.12}$$

$$T_1 = \left(\frac{1}{n_1} + \frac{1}{n_2}\right) \tag{8.13}$$

$$T_2 = \frac{(\Sigma x^2 - n_1\, m_x + \Sigma y^2 - n_2\, m_y)}{(n_1 + n_2 - 2)} \tag{8.14}$$

where m_x and m_y are the means for the x and y variables, d is the tested difference in the means, and n_1 and n_2 are the number of observations for variables x and y, respectively.

The Visual Basic Source Code

Let's look at the Visual Basic source code that implements the calculations for the basic statistics discussed in this chapter. Listing 8.1 contains the source code for the BASTAT.BAS module file.

 Throughout the book, the underscore character is used to split wrapping lines of Visual Basic declarations and statements.

Listing 8.1 The source code for the BASTAT.BAS module file.

```
Global Const BASTAT_EPS# = 1E-30

Type basicStat
 hasMissingData As Integer
 countData As Integer
 sum As Double
 sumX As Double
 sumX2 As Double
 sumX3 As Double
 sumX4 As Double
 missingCode As Double
 mean As Double
 sdev As Double
End Type

Function getMeanSdev (x() As Double, numData As Integer, _
 B As basicStat) As Integer

 Dim i As Integer
 Dim xx As Double, x2 As Double
```

Listing 8.1 (*Continued*)

```
If numData > 1 Then
   B.countData = B.countData + numData
   If B.hasMissingData Then
    For i = 0 To numData - 1
     xx = x(i)
     If xx > B.missingCode Then
       x2 = xx * xx
       B.sum = B.sum + 1
       B.sumX = B.sumX + xx
       B.sumX2 = B.sumX2 + x2
       B.sumX3 = B.sumX3 + xx * x2
       B.sumX4 = B.sumX4 + x2 * x2
     End If
    Next i
   Else
    For i = 0 To numData - 1
     xx = x(i)
     x2 = xx * xx
     B.sum = B.sum + 1
     B.sumX = B.sumX + xx
     B.sumX2 = B.sumX2 + x2
     B.sumX3 = B.sumX3 + xx * x2
     B.sumX4 = B.sumX4 + x2 * x2
    Next i
   End If
   If B.sum > 1 Then
    B.mean = B.sumX / B.sum
    B.sdev = Sqr((B.sumX2 - B.sumX ^ 2 / B.sum) / _
       (B.sum - 1#))
   Else
    getMeanSdev = False
    Exit Function
   End If
 End If
 getMeanSdev = False
End Function

Sub initializeBasicStat (B As basicStat, _
  hasMissingData As Integer, missingCode As Double)

 B.hasMissingData = hasMissingData
 B.missingCode = missingCode
 B.countData = 0
 B.sum = 0
 B.sumX = 0
 B.sumX2 = 0
 B.sumX3 = 0
 B.sumX4 = 0
 B.mean = 0
 B.sdev = 0
End Sub

Function meanCI (B As basicStat, probability As Double, _
  meanXLow As Double, meanXHi As Double) As Integer
' calculate the confidence interval of the mean value

 Dim tableT As Double, df As Double
 Dim delta As Double, p As Double

 If B.sum < 2 Then
   meanCI = False
   Exit Function
 End If
```

Listing 8.1 (*Continued*)

```
df = B.sum - 1#
If probability > 1# Then
  p = .5 - probability / 200#
Else
  p = .5 - probability / 2#
End If
tableT = TInv(p, df)
delta = tableT * B.sdev / Sqr(B.sum)
meanXHi = B.mean + delta
meanXLow = B.mean - delta
meanCI = True
End Function

Function meanTest (B1 As basicStat, B2 As basicStat, _
 testedDifference As Double, probability As Double, _
 calcT As Double, tableT As Double, passTest As Integer) _
 As Integer
' test means of two samples

 Dim df As Double, p As Double
 Dim factor1 As Double, factor2 As Double
 Dim n1 As Double, n2 As Double

 If B1.sum < 2 Or B2.sum < 2 Then
   meanTest = False
   Exit Function
 End If

 n1 = B1.sum
 n2 = B2.sum

 If probability > 1# Then
   p = .5 - probability / 200#
 Else
   p = .5 - probability / 2#
 End If
 df = n1 + n2 - 2#
 factor1 = Sqr(1# / n1 + 1# / n2)
 factor2 = Sqr((B1.sumX2 - B1.sumX ^ 2 / n1 + B2.sumX2 - _
     B2.sumX ^ 2 / n2) / df)
 calcT = (B1.sumX / n1 - B2.sumX / n2 - testedDifference) _
     / (factor1 * factor2)
 tableT = TInv(p, df)
 passTest = Abs(calcT) <= tableT
 meanTest = True
End Function

Function moments (B As basicStat, meanX As Double, _
 varianceX As Double, moment3 As Double, moment4 As Double, _
 skewnessCoeff As Double, kurtosisCoeff As Double) As Integer
' Function to calculate the first four moments and
'   the coefficients of skewness and kurtosis.

 Dim meanSqrd As Double

 If B.sum < 2 Then
   moments = False
   Exit Function
 End If

 meanX = B.sumX / B.sum
 meanSqrd = meanX ^ 2
 varianceX = B.sumX2 / B.sum - meanSqrd
```

Listing 8.1   (*Continued*)

```
moment3 = B.sumX3 / B.sum - 3# / B.sum * meanX * B.sumX2 + _
     2# * meanX * meanSqrd
moment4 = B.sumX4 / B.sum - 4# / B.sum * meanX * B.sumX3 + _
     6# / B.sum * meanSqrd * B.sumX2 - 3# * meanSqrd ^ 2
skewnessCoeff = moment3 / varianceX ^ 1.5
kurtosisCoeff = moment4 / varianceX ^ 2
moments = True
End Function


Function sdevCI (B As basicStat, probability As Double, _
sdevXLow As Double, sdevXHi As Double) As Integer
' calculate the confidence interval of the
'    standard deviation value

Dim df As Double, p As Double

If B.sum < 2 Then
  sdevCI = False
  Exit Function
End If
df = B.sum - 1#
If probability > 1# Then
  p = .5 - probability / 200#
Else
  p = .5 - probability / 2#
End If
sdevXLow = B.sdev * Sqr(df / ChiInv(p, df))
sdevXHi = B.sdev * Sqr(df / ChiInv((1 - p), df))
sdevCI = True
End Function
```

Listing 8.1 declares the data type basicStat. The structure contains fields that represent the statistical summations as well as the mean and standard deviation. In addition, the structure has the fields hasMissingData and missingCode to support missing data. The Visual Basic functions declared in the file use the basicStat structure to manage the data and statistical results.

The module file BASTAT.BAS declares the following Visual Basic functions:

1. The subroutine initializeBasicStat initializes a basicStat structure accessed by parameter r. The subroutine has two additional parameters that allow you to specify the status and value for missing data. You need to call InitializeBasicStat to initialize and reset the values in a basicStat structure.

2. The function getMeanSdev calculates the mean and standard deviation for data accessed by the array-parameter x. The parameter numData specifies how many elements of array x should be included in the calculations. You can repeatedly call getMeanSdev to process large number of observations in batches. This approach empowers you to use relatively small arrays to gradually read and process numerous observations. The function returns a Boolean value to indicate its success or failure.

3. The function meanCI calculates the confidence interval for the mean. This function has parameters that pass a basicStat structure, specify the confidence probability, and pass the low and high mean values back to the caller. The arguments for the latter two parameters must always be variables.

4. The function sdevCI calculates the confidence interval for the standard deviation. This function has parameters that pass a basicStat structure, specify the confidence probability, and pass the low and high standard deviation values back to the caller. The arguments for the latter two parameters must always be variables.

5. The function moments calculates the four moments and their related coefficients. This function has parameters that specify the basicStat structure (which provides the needed statistical summations) and the variables that report back the values of the moments and their coefficients. To use this function, you need to first call InitializeBasicStat and getMeanSdev (at least once).

6. The function meanTest performs the test for the mean of two samples. The function uses the parameters B1 and B2 as the the two basicStat structures, which already contain the summations and basic statistical values. The function has additional parameters that specify the tested mean difference, the test probability, the calculated Student-t value, the tabulated Student-t value, and the test outcome flag. The last three parameters report data back to the caller, so the arguments for these parameters must be variables.

The implementation of the Visual Basic functions in Listing 8.1 relies on using the functions file STATLIB.BAS to call the functions TInv and ChiInv, which return the inverse Student-t and Chi-square p.d.f. values, respectively.

Keep in mind that you need to call the functions in a certain sequence—you need to first initialize a basicStat structure, then accumulate data in that structure, before calling the functions that calculate the confidence interval, calculate the first moments, and test the sample means.

The Visual Basic Test Program

Let's look at the Visual Basic test program. Listing 8.2 shows the source code for the form in project file TSBASTAT.MAK. To compile the test program, you need to include the files BASTAT.BAS and STATLIB.BAS in your project file.

Listing 8.2 The source code for the form associated with the program project TSBASTAT.MAK.

```
Const SMALL_ARR# = 10
Const BIG_ARR1# = 50
Const BIG_ARR2# = 70

Dim B As basicStat

Sub CIMnu_Click ()
 Dim dummy As Integer
 Dim probab As Double
 Dim meanLow As Double
 Dim meanHi As Double
 Dim sdevLow As Double
 Dim sdevHi As Double

 probab = 95 ' 95%

 Cls
 Print "******** Confidence Interval ********"
 Print
```

Listing 8.2 (*Continued*)

```
 dummy = meanCI(B, probab, meanLow, meanHi)
 Print "At "; probab; "%, the confidence interval for ";
 Print "the mean is"
 Print Format$(meanLow, "##.##"); " to ";
 Print Format$(meanHi, "##.##")
 Print
 dummy = sdevCI(B, probab, sdevLow, sdevHi)
 Print "At "; probab; "%, the confidence interval for ";
 Print "the std. deviation is"
 Print Format$(sdevLow, "##.##"); " to ";
 Print Format$(sdevHi, "##.##")
End Sub

Sub ExitMnu_Click ()
 End
End Sub

Sub Form_Load ()
 CIMnu.Enabled = False
 FourMomentsMnu.Enabled = False
End Sub

Sub FourMomentsMnu_Click ()
 Dim dummy As Integer
 Dim meanX As Double
 Dim varX As Double
 Dim moment3 As Double
 Dim moment4 As Double
 Dim skewnessCoeff As Double
 Dim kurtosisCoeff As Double

 dummy = moments(B, meanX, varX, moment3, moment4, _
     skewnessCoeff, kurtosisCoeff)
 Cls
 Print "********* First Four Moments ***********"
 Print
 Print "Mean = "; Format$(meanX, "##.##")
 Print "Variance = "; Format$(varX, "###.##")
 Print "Moment 3 = "; Format$(moment3, "#.###E+##")
 Print "Moment 4 = "; Format$(moment4, "#.###E+##")
 Print "Skewness coeff. = "; Format$(skewnessCoeff, "#.###E+##")
 Print "Kurtosis coeff. = "; Format$(kurtosisCoeff, "#.###E+##")
End Sub

Sub MeanSdevMnu_Click ()
 Dim i As Integer
 Dim dummy As Integer
 Static x(SMALL_ARR) As Double

 ' initialize basic stat structure
 initializeBasicStat B, False, 0
 ' initialize array
 Randomize
 For i = 0 To SMALL_ARR - 1
   x(i) = 100 * Rnd(1)
 Next i
 ' get mean and sdev
 dummy = getMeanSdev(x(), SMALL_ARR, B)
 Cls
 Print "********* Mean and Std. Deviation ***********"
 Print
 Print "Mean = "; Format$(B.mean, "##.##")
 Print "Std. deviation = "; Format$(B.sdev, "##.##")
```

Listing 8.2 *(Continued)*

```
For i = 0 To SMALL_ARR - 1
  x(i) = 100 * Rnd(1)
Next i
' get mean and sdev
dummy = getMeanSdev(x(), SMALL_ARR, B)
Print
Print "After adding more data"
Print "Mean = "; Format$(B.mean, "##.##")
Print "Std. deviation = "; Format$(B.sdev, "##.##")

For i = 0 To SMALL_ARR - 1
  x(i) = 100 * Rnd(1)
Next i
' get mean and sdev
dummy = getMeanSdev(x(), SMALL_ARR, B)
Print
Print "After adding more data"
Print "Mean = "; Format$(B.mean, "##.##")
Print "Std. deviation = "; Format$(B.sdev, "##.##")
' enable the disabled menu selections
CIMnu.Enabled = True
FourMomentsMnu.Enabled = True
End Sub

Sub TestPairedMeansMnu_Click ()
Static x1(BIG_ARR2) As Double
Static x2(BIG_ARR2) As Double
Dim B1 As basicStat
Dim B2 As basicStat
Dim calcT As Double
Dim tableT As Double
Dim probab As Double
Dim passTest As Integer
Dim i As Integer
Dim dummy As Integer

' assign data to array x1
Randomize
For i = 0 To BIG_ARR2 - 1
  x1(i) = 100 * Rnd(1)
Next i

' assign data to array x2
Randomize
For i = 0 To BIG_ARR2 - 1
  x2(i) = 2000 * Rnd(1)
Next i
probab = 95

Cls
Print "**** Testing mean values of paired data ****"
Print
' initialize basic stat structure
initializeBasicStat B1, False, 0
initializeBasicStat B2, False, 0
' calc mean and sdev
dummy = getMeanSdev(x1(), BIG_ARR2, B1)
dummy = getMeanSdev(x2(), BIG_ARR2, B2)
dummy = meanTest(B1, B2, 0#, probab, calcT, tableT, passTest)
Print "At "; probab; "% the confidence"
Print "Calculated student-t = "; calcT
Print "Tabulated student-t = "; tableT
```

Listing 8.2 (*Continued*)

```
If passTest Then
   Print "Cannot reject that means are equal"
Else
   Print "Cannot reject that means are different"
End If
End Sub

Sub TestUnpairedMeansMnu_Click ()
Static x1(BIG_ARR1) As Double
Static x2(BIG_ARR2) As Double
Dim B1 As basicStat
Dim B2 As basicStat
Dim calcT As Double
Dim tableT As Double
Dim probab As Double
Dim passTest As Integer
Dim i As Integer
Dim dummy As Integer

' assign data to array x1
Randomize
For i = 0 To BIG_ARR1 - 1
   x1(i) = 100 * Rnd(1)
Next i

' assign data to array x2
Randomize
For i = 0 To BIG_ARR2 - 1
   x2(i) = 100 * Rnd(1)
Next i
probab = 95

Cls
Print "**** Testing mean values of unpaired data ****"
Print
' initialize basic stat structures
initializeBasicStat B1, False, 0
initializeBasicStat B2, False, 0
  ' calc mean and sdev
dummy = getMeanSdev(x1(), BIG_ARR1, B1)
dummy = getMeanSdev(x2(), BIG_ARR2, B2)
dummy = meanTest(B1, B2, 0#, probab, calcT, tableT, passTest)
Print "At "; probab; "% the confidence"
Print "Calculated student-t = "; calcT
Print "Tabulated student-t = "; tableT
If passTest Then
   Print "Cannot reject that means are equal"
Else
   Print "Cannot reject that means are different"
End If
End Sub
```

The project uses a form that has a simple menu system but no controls. Table 8.1 shows the menu structure and the names of the menu items. The form has the caption "Basic Statistics." The menu option Test has five selections to test the various basic statisics. Each one of these menu selections clears the form and then displays the result of some basic statisical calculations. Thus, you can zoom in on any method by invoking its related menu selection.

**TABLE 8.1 The Menu System for the
TSBASTAT.MAK Project**

Menu caption	Name
&Test	TesMnu
Mean and Std. Deviation	MeanSdevMnu
Confidence Interval	CIMnu
Four Moments	FourMomentsMnu
Test Paired Means	TestPairedMeansMnu
Test Unpaired Means	TestUnpairedMeansMnu
–	N1
&Exit	ExitMnu

The program tests the various basic statistics functions using internal data. To test the basic statistics, the program performs the following tasks:

1. Declares the basicStat-type structure B on the form level to be accessed by the event-handlers of the first three menu selections

2. Disables the second and third menu selections when the form loads

The subroutine MeanSdevMnu performs the following tasks:

1. Initializes the structure B by calling the function initializeBasicStat

2. Stores random numbers (in the range of 0 to 99) in the local array x

3. Calculates and displays the mean and standard deviation values for the array x by calling the function getMeanSdev with the arguments x, SMALL_ARR, and the structure B (This task displays the sought statistics by accessing the fields mean and sdev of the structure B.)

4. Repeats steps 2 and 3 twice to add more data to the fields of structure B

5. Enables the second and third menu selections (You can now use these menu selections, since the form-level structure B contains valid data.)

The subroutine CIMnu performs the following tasks:

1. Calculates and displays the confidence interval for the mean value, involving a call to the function meanCI

2. Calculates and displays the confidence interval for the !SD value, involving a call to the function sdevCI

The subroutine FourMomentsMnu calculates and displays the first four moments and their related coefficients. This task involves calling the function moments.

The subroutine TestUnpairedMeanMnu performs the following tasks:

1. Stores an unequal number of random numbers in the local arrays x1 and x2 (Both arrays contain random numbers in the range of 0 to 99.)

2. Tests the means of arrays x1 and x2 (This task involves initializing the structures B1 and B2, followed by adding data using function getMeanSdev, and then finally calling function meanTest to test the means of the unpaired samples. The program displays the outcome of the statistical test of the two means)

The subroutine TestPairedMeanMnu performs the following tasks:

1. Stores random numbers in the arrays x1 and x2 (The array x1 contains random numbers in the range of 0 to 99, whereas the array x2 contains random numbers in the range of 0 to 1999.)

2. Tests the means of arrays x1 and x2, initializing the structures B1 and B2, followed by adding data using function getMeanSdev, and then finally calling function meanTest to test the means of the paired samples (The program displays the outcome of the statistical test of the two means.)

Figure 8.1 shows the output of the test program for each menu option.

```
********* Mean and Std. Deviation ***********

Mean = 46.96
Std. deviation = 29.01

After adding more data
Mean = 46.66
Std. deviation = 34.73

After adding more data
Mean = 47.75
Std. deviation = 33.25


******** Confidence Interval ********

At 95 %, the confidence interval for the mean is
35.35 to 60.15

At 95 %, the confidence interval for the std. deviation is
26.49 to 44.69


********* First Four Moments ***********

Mean = 47.75
Variance = 1068.63
Moment 3 = 3.133E+3
Moment 4 = 1.747E+6
Skewness coeff. = 8.968E-2
Kurtosis coeff. = 1.53E+0
```

Figure 8.1 The output of the test program for basic statistics.

```
**** Testing mean values of paired data ****

At 95 % the confidence
Calculated student-t = -14.7278656004349
Tabulated student-t = 1.97452147550324
Cannot reject that means are different


**** Testing mean values of unpaired data ****

At 95 % the confidence
Calculated student-t = 1.52341201248917
Tabulated student-t = 1.97747942943271
Cannot reject that means are equal
```

Figure 8.1 (*Continued*)

Chapter

9

The ANOVA Tests

Statistics offer several kinds of analysis of variance (ANOVA) tests that enable you to determine whether or not changes in data are statistically significant. This chapter provides the methods and Visual Basic source code for the following ANOVA tests:

- The one-way ANOVA

- The two-way ANOVA with no replications

- The two-way ANOVA with replications

- The Latin-square ANOVA

- The analysis of covariance ANOCOV

The One-Way ANOVA

The one-way ANOVA tests whether the observed differences among a set of sample means can be explained by chance or whether these variations indicate the actual differences among the corresponding population means. The null hypothesis for the one-way ANOVA is that the population means are all equal. The one-way ANOVA involves calculating several items:

- The mean value for each of the N samples:

$$m_i = \frac{\Sigma x_{ij}}{n_i} \text{ for i = 1 to N and j = 1 to } n_i \tag{9.1}$$

- The standard deviation for each of the N samples:

$$s_i = \Sigma \left(\frac{(\Sigma x_{ij}^2 - n_i m_i^2)}{(n_i - 1)} \right) \tag{9.2}$$

- The sum of observations for each of the N samples:

$$\text{Sum}_i = \Sigma x_{ij} \tag{9.3}$$

- The sum of squares:

$$\text{TSS} = \frac{\Sigma\Sigma x_{ij}^2 - (\Sigma\Sigma x_{ij})^2}{\Sigma n_i} \tag{9.4}$$

- The treatment sum of squares:

$$\text{TrSS} = \frac{\Sigma[(\Sigma x_{ij})^2 / n_i] - (\Sigma\Sigma x_{ij})^2}{\Sigma n_i} \tag{9.5}$$

- The error sum of squares:

$$\text{ESS} = \text{TSS} - \text{TrSS} \tag{9.6}$$

- The treatment degrees of freedom:

$$df_1 = N - 1 \tag{9.7}$$

- The error degrees of freedom:

$$df_2 = \Sigma n_i - N \tag{9.8}$$

- The total degrees of freedom:

$$df_3 = df_1 + df_2 \tag{9.9}$$

- The treatment mean square:

$$\text{TrMS} = \frac{\text{TrSS}}{df_1} \tag{9.10}$$

- The error mean square:

$$\text{EMS} = \frac{\text{ESS}}{df_2} \tag{9.11}$$

- The F ratio:

$$F = \frac{\text{TrMS}}{\text{EMS}} \tag{9.12}$$

These statistics are typically arranged and displayed as follows:

	SS	df	MS	F
Treatments	TrSS	df_1	TrMS	F
Error	ESS	df_2	EMS	
Total	TSS	df_3		

Here is the algorithm for the one-way ANOVA:

Given:

- The matrix Xmat, which contains N columns and M rows

Algorithm:

1. For I = 0 to N – 1, set Sum[I] = 0, SumX[I] = 0, and SumX2[I] = 0
2. For I = 0 to M – 1, repeat the following:
   2.1. For J = 0 to N – 1, repeat the following:
       2.1.1. Increment Sum[J]
       2.1.2. Add Xmat[I, J] to SumX[J]
       2.1.3. Add Xmat[I, J] squared to SumX2[J]
3. Set TS = 0, TSS = 0, and GrandTotal = 0
4. For I = 0 to N – 1, repeat the following steps:
   4.1. Set $\text{Mean[I]} = \dfrac{\text{SumX[I]}}{\text{Sum[I]}}$

   4.2. Set $\text{Sdev[i]} = \sqrt{\left(\dfrac{\dfrac{(\text{SumX2[I]} - \text{SumX[I]}\^2)}{(\text{Sum[i]})}}{(\text{Sum[i]} - 1)} \right)}$

   4.3. Add SumX[I] to TS
   4.4. Add SumX2[I] to TSS
   4.5. Add Sum[I] to GrandTotal
5. Set $\text{TSS} = \dfrac{\text{TSS} - \text{TS}\^2}{\text{GrandTotal}}$
6. Set TrSS = 0
7. For I = 0 to N – 1, add $\dfrac{\text{SumX[I]}\^2}{\text{Sum[I]}}$ to TrSS
8. Set $\text{TrSS} = \dfrac{\text{TrSS} - \text{TS}\^2}{\text{GrandTotal}}$
9. Set ESS = TSS – TrSS
10. Set df1 = N – 1
11. Set df2 = GrandTotal – df1 – 1
12. Set df3 = df1 + df2
13. Set $\text{TrMS} = \dfrac{\text{TrSS}}{\text{df1}}$
14. Set $\text{EMS} = \dfrac{\text{ESS}}{\text{df2}}$
15. Set $\text{F} = \dfrac{\text{TrMS}}{\text{EMS}}$

The Two-Way ANOVA

The two-way ANOVA with no replications analyzes the total variability of a set of data into components that can be attributed to diverse sources of variation. The two-way ANOVA examines the row effects and the column effects independently.

The two-way ANOVA requires calculating several statistics given by the following equations:

- The column and row sums:

$$RS_i = \Sigma x_{ij} \text{ for } i = 1 \text{ to } r$$

$$CS_i = \Sigma x_{ij} \text{ for } j = 1 \text{ to } c$$

- The sums of squares for the rows and columns:

$$RSS = \frac{\Sigma (\Sigma x_{ij})^2}{c} - \frac{(\Sigma \Sigma x_{ij})^2}{rc}$$

$$CSS = \frac{\Sigma (\Sigma x_{ij})^2}{r} - \frac{(\Sigma \Sigma x_{ij})^2}{rc}$$

- The total sum of squares:

$$TSS = \frac{\Sigma \Sigma x_{ij}^2 - (\Sigma \Sigma x_{ij})^2}{rc}$$

- The error sum of squares:

$$ESS = TSS - RSS - CSS$$

- The degrees of freedom for the row effect, the column effect, and the error:

$$df_1 = r - 1$$

$$df_2 = c - 1$$

$$df_3 = (r - 1)(c - 1)$$

- The F ratios for the row and column:

$$F_1 = \frac{\left(\dfrac{RSS}{df_1}\right)}{\left(\dfrac{ESS}{df_3}\right)}$$

$$F_2 = \frac{\left(\dfrac{CSS}{df_2}\right)}{\left(\dfrac{ESS}{df_3}\right)}$$

These statistics are typically arranged and displayed as follows:

The ANOVA Tests

	SS	df	F
Row	RSS	df_1	F_1
Column	CSS	df_2	F_2
Error	ESS	df_3	
Total	TSS		

Here is the algorithm for the two-way ANOVA with no replication:

Given:

- The matrix Xmat with R rows and C columns

Algorithm:

1. For I = 0 to C − 1, repeat the following steps:
   1.1. Set ColSumX[I] = 0
   1.2. Set ColSumX2[I] = 0

2. For I = 0 to R − 1, repeat the following steps:
   2.1. Set RowSumX[I] = 0
   2.2. Set RowSumX2[I] = 0

3. Set GrandSumX = 0

4. Set GrandSumX2 = 0

5. For I = 0 to R − 1, repeat the next steps:
   5.1. For J = 0 to C − 1, repeat the next steps:
      5.1.1. Add Xmat[I, J] to RowSumX[I]
      5.1.2. Add Xmat[I, J] to ColSumX[J]
      5.1.3. Add Xmat[I, J] squared to RowSumX2[I]
      5.1.4. Add Xmat[I, J] squared to ColSumX2[J]

6. For I = 0 to R − 1, repeat the next steps:
   6.1. Add RowSumX[I] to GrandSumX
   6.2. Add RowSumX2[I] to GrandSumX2

7. Set RC = R * C

8. Set TSS = $\dfrac{\text{GrandSumX2} - \text{GrandSumX}^2}{RC}$

9. Set RSS = 0

10. For I = 0 to R − 1, add $\dfrac{\text{RowSumX[I]}^2}{C}$ to RSS

11. Subtract $\dfrac{\text{GrandSumX}^\wedge 2}{\text{RC}}$ from RSS

12. Set CSS = 0

13. For I = 0 to C − 1, add $\dfrac{\text{GrandSumX}^\wedge 2}{\text{R}}$ to CSS

14. Subtract $\dfrac{\text{GrandSumX}^\wedge 2}{\text{RC}}$ from CSS

15. Set ESS = TSS − RSS − CSS

16. Set RowDf = R − 1

17. Set ColDf = C − 1

18. Set ErrDf = RowDf ∗ ColDf

19. Set TotalDf = RC − 1

20. Set F1 = $\dfrac{\left(\dfrac{\text{RSS}}{\text{RowDf}}\right)}{\left(\dfrac{\text{ESS}}{\text{ErrDf}}\right)}$

20. Set F2 = $\dfrac{\left(\dfrac{\text{CSS}}{\text{ColDf}}\right)}{\left(\dfrac{\text{ESS}}{\text{ErrDf}}\right)}$

The Two-Way ANOVA with Replication

The two-way ANOVA with replication allows you to have multiple observations for each row and column effect. The multiple samples allow this kind of ANOVA test to study the interaction effect that might appear due to the unique combination of the row and column effects.

The two-way ANOVA requires calculating several statistics, given by the following equations:

- The cell summations matrix:

$$LS_{ij} = \Sigma x_{ijk} \text{ for k = 1 to number of replicates m}$$

- The row and column sums:

$$RS_i = \Sigma\Sigma x_{ijk} \text{ for k = 1 to m, j = 1 to c}$$

$$CS_i = \Sigma\Sigma x_{ijk} \text{ for k = 1 to m, i = 1 to r}$$

- The grand sum and sum of squares:

$$G = \Sigma\Sigma\Sigma x_{ijk}$$

$$TSS = \Sigma\Sigma\Sigma x_{ikj}{}^2$$

- The sums of squares for the rows, columns, interactions, and error:

$$RSS = \Sigma\left(\frac{RS_i^2}{rm}\right) - \frac{G^2}{rcm}$$

$$CSS = \Sigma\left(\frac{CS_i^2}{cm}\right) - \frac{G^2}{rcm}$$

$$ISS = \Sigma\Sigma\left(\frac{LS_{ij}^2}{m}\right) - \frac{G^2}{rcm} - RSS - CSS$$

$$ESS = SS - \Sigma\Sigma\left(\frac{LS_{ij}^2}{m}\right)$$

- The degrees of freedom for the row effect, column effect, interaction, and error:

$$df_1 = r - 1$$
$$df_2 = c - 1$$
$$df_3 = (r - 1)(c - 1)$$
$$df_4 = rc(m - 1)$$

- The F ratios for the row, column, and interaction:

$$F_1 = \frac{\left(\dfrac{RSS}{df_1}\right)}{\left(\dfrac{ESS}{df_4}\right)}$$

$$F_2 = \frac{\left(\dfrac{CSS}{df_2}\right)}{\left(\dfrac{ESS}{df_4}\right)}$$

$$F_3 = \frac{\left(\dfrac{ISS}{df_3}\right)}{\left(\dfrac{ESS}{df_4}\right)}$$

These statistics are typically arranged and displayed as follows:

	SS	**df**	**F**
Row	RSS	df_1	F_1
Column	CSS	df_2	F_2
Interaction	ISS	df_3	F_3
Error	ESS	df_4	
Total	TSS		

Here is the algorithm for the two-way ANOVA with replication:

Given:

- The data matrix Xmat with R rows, C columns, and M replicates

Algorithm:

1. For I = 0 to R – 1, set RowSumX[I] = 0

2. For I = 0 to C – 1, set ColSumX[I] = 0

3. For I = 0 to R – 1 repeat the next step:
   3.1. For J = 0 to C – 1 repeat the next step:
       3.1.1. Set CellSum[I, J] = 0

4. Set GramdSumX = 0

5. Set GrandSumX2 = 0

6. For I = 0 to R – 1, repeat the next step:
   6.1. Set rowIndex = $\dfrac{i}{M}$
   6.2. For J = 0 to C – 1, repeat the next step:
       6.2.1. Add X[I, J] to CellSum[rowIndex, J] = 0
       6.2.2. Add X[I, J] to RowSumX[rowIndex]
       6.2.3. Add X[I, J] to ColSumX[J]
       6.2.4. Add X[I, J] squared to GrandSumX2

7. Set K = $\dfrac{R}{M}$

8. Set NRow = K and NCol = C

9. For I = 0 to C – 1, add ColSumX[I] to GrandSumX

10. Set RC = NRow * NCol * M

11. Set TSS = $\dfrac{\text{GrandSumX2} - \text{GrandSumX\textasciicircum 2}}{\text{RC}}$

12. Set RSS = 0

13. For I = 0 to NRow – 1, add $\dfrac{\text{RowSumX[I]\textasciicircum 2}}{(\text{NCol} * \text{M})}$ to RSS

14. Subtract $\dfrac{\text{GrandSumX\textasciicircum 2}}{\text{RC}}$ from RSS

15. Set CSS = 0

16. For I = 0 to NCol – 1, add $\dfrac{\text{ColSumX[I]\textasciicircum 2}}{(\text{NRow} * \text{M})}$ to CSS

17. Subtract $\dfrac{\text{GrandSumX\textasciicircum 2}}{\text{RC}}$ from CSS

18. Set Sum = 0

19. For I = 0 to NRow, repeat the following steps:

    19.1. For J = 0 to NCol, repeat the following step:

        19.1.1. Add CellSum[I, J]^2 to Sum

20. Set ISS = $\dfrac{\text{Sum}}{M}$ – RSS – CSS – $\dfrac{\text{GrandSumX}^{\wedge}2}{RC}$

21. Set ESS = GrandSumX2 – $\dfrac{\text{Sum}}{M}$

22. Set TSS = $\dfrac{\text{GrandSumX2} - \text{GrandSumX}^{\wedge}2}{RC}$

23. Set RowDf = NRow – 1

24. Set ColDf = NCol – 1

25. Set InteractDf = RowDf * ColDf

26. Set TotalDf = RC – 1

27. Set ErrorDf = NRow * NCol * (M – 1)

28. Set RowF = $\left(\dfrac{\text{RSS}}{\text{ESS}}\right) * \left(\dfrac{\text{ErrorDf}}{\text{RowDf}}\right)$

29. Set ColF = $\left(\dfrac{\text{CSS}}{\text{ESS}}\right) * \left(\dfrac{\text{ErrorDf}}{\text{ColDf}}\right)$

30. Set InteractF = $\left(\dfrac{\text{ISS}}{\text{ESS}}\right) * \left(\dfrac{\text{ErrorDf}}{\text{InteractDf}}\right)$

The Latin-Square ANOVA

The Latin-square ANOVA tests for the effects of three factors, A, B, and C. The Latin-square ANOVA uses a data table with an equal number of rows and columns. Here is an example of a 5-by-5 Latin square experimental design:

	B_1	B_2	B_3	B_4	B_5
A_1	C_1	C_2	C_3	C_4	C_5
A_2	C_4	C_5	C_1	C_2	C_3
A_3	C_5	C_1	C_2	C_3	C_4
A_4	C_2	C_3	C_4	C_5	C_1
A_5	C_3	C_4	C_5	C_1	C_2

You can call the effect A the *row effect*, the effect B the *column effect*, and the effect C the *treatment effect*. The ANOVA table calculates the following statistics:

- The sums of effects A, B, and C:

$$SA_i = \Sigma x_{ij} \quad \text{for } j = 1 \text{ to } r$$

$$SB_i = \Sigma x_{ij} \quad \text{for } j = 1 \text{ to } r$$

$$SC_i = \Sigma x_{ij}$$

- The grand sum and sum of square:

$$G = \Sigma\Sigma x_{ij}$$

$$TSS = \Sigma\Sigma x_{ij}^2$$

- The sum of squares for the three effects, the sum of squares for the error, and the total sum of squares:

$$TSA = \Sigma\left(\frac{SA_i^2}{r}\right) - \frac{G^2}{r^2}$$

$$TSB = \Sigma\left(\frac{SB_i^2}{r}\right) - \frac{G^2}{r^2}$$

$$TSC = \Sigma\left(\frac{SC_i^2}{r}\right) - \frac{G^2}{r^2}$$

$$ESS = TSS - TSA - TSB - TSC$$

$$TSS = SS - \frac{G^2}{r^2}$$

- The degrees of freedom for the effects and for the error:

$$Adf = r - 1$$

$$Bdf = r - 1$$

$$Cdf = r - 1$$

$$Errdf = (r - 1)(r - 2)$$

- The F ratios for the three effects:

$$FA = \frac{(r-2)TSA}{ESS}$$

$$FB = \frac{(r-2)TSB}{ESS}$$

$$FC = \frac{(r-2)TSC}{ESS}$$

These statistics are typically arranged and displayed as follows:

Effect	SS	df	F
A	TSA	Adf	FA
B	TSB	Bdf	FB
C	TSC	Cdf	FC
Error	ESS	ErrDf	
Total	TSS		

Here is the algorithm for the Latin-square ANOVA:

Given:

- The matrix Xmat with R rows and columns
- The indexing matrix Map, which specifies how the factor C appears in the matrix Xmat

Algorithm:

1. For I = 0 to R − 1, repeat the following steps:
   1.1. Set ColSumX[I] = 0
   1.2. Set RowSumX[I] = 0
   1.3. Set TrtSumX[I] = 0
2. Set GrandSumX = 0
3. Set GrandSumX2 = 0
4. Set N = 0
5. For I = 0 to R − 1, repeat the following steps:
   5.1. Increment N
   5.2. For J = 0 to R − 1, repeat the following steps:
       5.2.1. Add Xmat[I, J] to ColSumX[J]
       5.2.2. Add Xmat[I, J] to RowSumX[I]
       5.2.3. Set K = Map[I, J]
       5.2.4. Add X to TrtSumX[K]
       5.2.5. Add Xmat[I, J] squared to GrandSumX2
   5.3. Add RowSumX[I] to GrandSumX
6. Set RSS = 0, CSS = 0, and TrtSS = 0
7. For I = 0 to R − 1, repeat the following steps:
   7.1. Add RowSumX[I] squared to RSS
   7.2. Add ColSumX[I] squared to CSS
   7.3. Add TrtSumX[I] squared to TrtSS
8. Set $G = \dfrac{\text{GrandSumX}}{N}$
9. Set $RSS = \dfrac{\text{RSS}}{N} - G$
10. Set $CSS = \dfrac{\text{CSS}}{N} - G$
11. Set $TrtSS = \dfrac{\text{TrtSS}}{N} - G$
12. Set ESS = GrandSumX2 − RSS − CSS − TrtSS − G
13. Set TSS = GrandSumX2 − G
14. Set RowDf = N − 1

15. Set ColDf = N − 1

16. Set TrtDf = N − 1

17. Set ErrorDf = (N − 1) * (N − 2)

18. Set TotalDf = N * N − 1

19. Set RowF = (N − 2) * $\dfrac{RSS}{ESS}$

20. Set ColF = (N − 2) * $\dfrac{CSS}{ESS}$

21. Set TrtF = (N − 2) * $\dfrac{TrtSS}{ESS}$

The Analysis of Covariance

The analysis of covariance (ANOCOV) examines the effects of variable separately from the effect of a second variable. The test assumes that the second variable can be measured. The analysis of covariance assumes that there are k populations, each with n_i number of observations. Thus, each population may have a varying number of observations.

The ANOCOV statistics calculate the following items:

- The sums and sums of squares:

$$Sx_i = \Sigma x_{ij} \quad \text{for } i = 1 \text{ to } k$$

$$TSSx = \frac{\Sigma\Sigma x_{ij}{}^2 - (\Sigma\Sigma x_{ij})^2}{\Sigma n_i}$$

$$ASSx = \Sigma\left[\frac{(\Sigma x_{ij})^2}{n_i}\right] - \frac{(\Sigma\Sigma x_{ij})^2}{\Sigma n_i}$$

$$WSSx = TSSx - ASSx$$

- The degrees of freedoms:

$$df_1 = k - 1$$

$$df_2 = \Sigma n_i - k$$

- The mean squares and F statistic

$$AMSx = \frac{ASSx}{df_1}$$

$$WMSx = \frac{WSSx}{df_2}$$

$$Fx = \frac{AMSx}{WMSx}$$

You can generate the preceding items for the variable y by replacing x_{ij} with y_{ij}.

- The sum of products:

$$TSP = \frac{\Sigma\Sigma x_{ij}\, y_{ij} - (\Sigma x_{ij})(\Sigma y_{ij})}{\Sigma n_i}$$

$$ASP = \Sigma\left[\frac{(\Sigma x_{ij})(\Sigma y_{ij})}{n_i}\right] - \frac{(\Sigma\Sigma x_{ij})(\Sigma\Sigma y_{ij})}{\Sigma n_i}$$

$$WSP = TSP - ASP$$

- The residual sums of squares:

$$TSSy^\wedge = \frac{TSSy - TSP^\wedge 2}{TSSx}$$

$$WSSy^\wedge = \frac{WSSy - WSP^\wedge 2}{WSSx}$$

$$ASSy^\wedge = TSSy^\wedge - WSSy^\wedge$$

- The residual degrees of freedom:

$$df_3 = k - 1$$

$$df_4 = \Sigma n_i - k - 1$$

- The residual mean squares and F statistics:

$$AMSy^\wedge = \frac{ASSy^\wedge}{df_3}$$

$$WMSy^\wedge = \frac{WSSy^\wedge}{df_4}$$

$$F = \frac{AMSy^\wedge}{WMSy^\wedge}$$

Here is the algorithm for ANOCOV:

Given:

- The matrix Xmat with K sets of X and Y pairs of observations
- N[I] observations in each set
- The maximum number of observations in any set, NData

Algorithm:

1. Set N = 2 * K * NData
2. For i = 0 to N − 1, set Sum[I], SumX[I], SumX2[I], SumY[I], SumY2[I], and SumXY[I] to 0
3. Set GrandSum = 0
4. Set ASSx = 0

5. Set ASSy = 0

6. Set ASP = 0

7. For I = 0 to NData − 1, repeat the following:
   7.1. For J = 0 to K − 1, repeat the following:
      7.1.1. Set jY = 2 ∗ J + 1
      7.1.2. Set jX = jY − 1
      7.1.3. If Xmat[I, J] is not missing, perform the next steps:
         7.1.3.1. Set x = Xmat[I, jX]
         7.1.3.2. Set y = Xmat[I, jY]
         7.1.3.3. Increment Sum[J]
         7.1.3.4. Add x to SumX[J]
         7.1.3.5. Add x squared to SumX2[J]
         7.1.3.6. Add y to SumY[J]
         7.1.3.7. Add y squared to SumY2[J]
         7.1.3.8. Add x ∗ y to SumXY[I]

8. Set dSumX, dSumY, dSumX2, dSumY2, dSumXY = 0

9. For J = 0 to K − 1, repeat the next steps:

   9.1. Add $\dfrac{\text{SumX[J]\^2}}{\text{Sum[J]}}$ to ASSx

   9.2. Add $\dfrac{\text{SumY[J]\^2}}{\text{Sum[J]}}$ to ASSy

   9.3. Add $\dfrac{\text{SumX[J]} \ast \text{SumY[J]}}{\text{Sum[J]}}$ to ASP

   9.4. Add Sum[J] to GrandSum
   9.5. Add SumX[J] to dSumX
   9.6. Add SumX2[J] to dSumX2
   9.7. Add SumY[J] to dSumY
   9.8. Add SumY2[J] to dSumY2
   9.9. Add SumXY[J] to dSumXY

10. Set TSSx = $\dfrac{\text{dSumX2} - \text{dSumX\^2}}{\text{GrandSum}}$

11. Set TSSy = $\dfrac{\text{dSumY2} - \text{dSumY\^2}}{\text{GrandSum}}$

12. Set ASSx = $\dfrac{\text{ASSx} - \text{dSumX\^2}}{\text{GrandSum}}$

13. Set ASSy = $\dfrac{\text{ASSy} - \text{dSumY\^2}}{\text{GrandSum}}$

14. Set WSSx = TSSx − ASSx

15. Set WSSy = TSSy − ASSy

16. Set Df1 = K − 1

17. Set Df3 = Df1

18. Set Df2 = GrandSum − K

19. Set Df4 = Df2 − 1

20. Set TSP = $\dfrac{\text{dSumXY} - \text{dSumX} * \text{dSumY}}{\text{GrandSum}}$

21. Set ASP = $\dfrac{\text{dSumX} * \text{dSumY}}{\text{GrandSum}}$

22. Set WSP = TSP − ASP

23. Set AMSx = $\dfrac{\text{ASSx}}{\text{Df1}}$

24. Set AMSy = $\dfrac{\text{ASSy}}{\text{Df1}}$

25. Set WMSx = $\dfrac{\text{WSSx}}{\text{Df2}}$

26. Set WMSy = $\dfrac{\text{WSSy}}{\text{Df2}}$

27. Set Fx = $\dfrac{\text{AMSx}}{\text{WMSx}}$

28. Set Fy = $\dfrac{\text{AMSy}}{\text{WMSy}}$

29. Set TSSyhat = $\dfrac{\text{TSSy} - \text{TSP}^2}{\text{TSSx}}$

30. Set WSSyhat = $\dfrac{\text{WSSy} - \text{WSP}^2}{\text{WSSx}}$

31. Set ASSyhat = TSSyhat − WSSyhat

32. Set AMSyhat = $\dfrac{\text{ASSyhat}}{\text{Df3}}$

33. Set WMSyhat = $\dfrac{\text{WSSyhat}}{\text{Df4}}$

34. Set F = $\dfrac{\text{AMSyhat}}{\text{WMSyhat}}$

The Visual Basic Source Code

Let's look at the Visual Basic source code that implements the various kinds of ANOVA statistics presented in this chapter. Listing 9.1 shows the source code for the ANOVA.BAS module file.

 Throughout the book, the underscore character is used to split wrapping lines of Visual Basic declarations and statements.

Listing 9.1   The source code for the ANOVA.BAS module file.

```
Global Const ANOVA_EPS# = 1E-30

Type ANOVA1rec
 NVar As Integer
 Treatment_SS As Double
 Treatment_df As Double
 Treatment_MS As Double
 Error_SS As Double
 Error_df As Double
 Error_MS As Double
 Total_SS As Double
 Total_df As Double
 ANOVA1_F As Double
 hasMissingData As Integer
 missingCode As Double
End Type

Type ANOVA2rec
 Num_Row As Integer
 Num_Col As Integer
 GrandSumX As Double
 GrandSumX2 As Double
 Row_SS As Double
 Row_df As Double
 Row_F As Double
 Col_SS As Double
 Col_df As Double
 Col_F As Double
 Error_SS As Double
 Error_df As Double
 Total_SS As Double
 Total_df As Double
End Type

Type ANOVA2Rrec
 NReplicate As Integer
 Num_Row As Integer
 Num_Col As Integer
 GrandSumX As Double
 GrandSumX2 As Double
 Row_SS As Double
 Row_df As Double
 Row_F As Double
 Col_SS As Double
 Col_df As Double
 Col_F As Double
 Interact_SS As Double
 Interact_df As Double
 Interact_F As Double
 Error_SS As Double
 Error_df As Double
 Total_SS As Double
 Total_df As Double
End Type

Type LatinSqrrec
 N As Integer
 Num As Double
 GrandSumX As Double
 GrandSumX2 As Double
 Row_SS As Double
 Row_df As Double
 Row_F As Double
```

Listing 9.1 (*Continued*)

```
 Col_SS As Double
 Col_df As Double
 Col_F As Double
 Trt_SS As Double
 Trt_df As Double
 Trt_F As Double
 Error_SS As Double
 Error_df As Double
 Total_SS As Double
 Total_df As Double
End Type

Type ANOCOVrec
 Num_Set As Integer
 hasMissingData As Integer
 missingCode As Double
 GrandSum As Double
 Df1 As Double
 Df2 As Double
 Df3 As Double
 Df4 As Double
 ASSx As Double
 ASP As Double
 ASSy As Double
 ASSyhat As Double
 AMSyhat As Double
 WSSx As Double
 WSP As Double
 WSSy As Double
 WSSyhat As Double
 WMSyhat As Double
 TSSx As Double
 TSP As Double
 TSSy As Double
 TSSyhat As Double
 ANOCOV_F As Double
 AMSx As Double
 WMSx As Double
 AMSy As Double
 WMSy As Double
 Fx As Double
 Fy As Double
End Type

Sub ANOCOV (r As ANOCOVrec, DataMat() As Double, _
 NData As Integer, NumSets As Integer, _
 hasMissingData As Integer, missingCode As Double, _
 sum() As Double, sumX() As Double, sumY() As Double, _
 sumX2() As Double, sumY2() As Double, sumXY() As Double)
 Dim i As Integer, j As Integer
 Dim jX As Integer, jY As Integer
 Dim x As Double, y As Double
 Dim dSumX As Double, dSumY As Double
 Dim dSumX2 As Double, dSumY2 As Double
 Dim dSumXY As Double
 Dim N As Integer

 N = 2 * NumSets

 ' initialize statistical summations
 For i = 0 To N - 1
   sum(i) = 0
   sumX(i) = 0
```

Listing 9.1 (*Continued*)

```
   sumX2(i) = 0
   sumY(i) = 0
   sumY2(i) = 0
   sumXY(i) = 0
Next i
r.GrandSum = 0
r.ASSx = 0
r.ASSy = 0
r.ASP = 0
r.hasMissingData = hasMissingData
r.missingCode = missingCode

' update statistical summations
r.Num_Set = NumSets
For i = 0 To NData - 1
  For j = 0 To r.Num_Set - 1
    jY = 2 * j + 1
    jX = jY - 1
    x = DataMat(i, jX)
    If Not (r.hasMissingData And x <= r.missingCode) Then
     y = DataMat(i, jY)
     sum(j) = sum(j) + 1
     sumX(j) = sumX(j) + x
     sumX2(j) = sumX2(j) + x ^ 2
     sumY(j) = sumY(j) + y
     sumY2(j) = sumY2(j) + y ^ 2
     sumXY(j) = sumXY(j) + x * y
    End If
  Next j
Next i
' carry out ANOCOV calculations
dSumX = 0
dSumY = 0
dSumX2 = 0
dSumY2 = 0
dSumXY = 0
For j = 0 To r.Num_Set - 1
  r.ASSx = r.ASSx + sumX(j) ^ 2 / sum(j)
  r.ASSy = r.ASSy + sumY(j) ^ 2 / sum(j)
  r.ASP = r.ASP + sumX(j) * sumY(j) / sum(j)
  r.GrandSum = r.GrandSum + sum(j)
  dSumX = dSumX + sumX(j)
  dSumX2 = dSumX2 + sumX2(j)
  dSumY = dSumY + sumY(j)
  dSumY2 = dSumY2 + sumY2(j)
  dSumXY = dSumXY + sumXY(j)
Next j
r.TSSx = dSumX2 - dSumX ^ 2 / r.GrandSum
r.TSSy = dSumY2 - dSumY ^ 2 / r.GrandSum
r.ASSx = r.ASSx - dSumX ^ 2 / r.GrandSum
r.ASSy = r.ASSy - dSumY ^ 2 / r.GrandSum
r.WSSx = r.TSSx - r.ASSx
r.WSSy = r.TSSy - r.ASSy
r.Df1 = r.Num_Set - 1
r.Df3 = r.Df1
r.Df2 = r.GrandSum - r.Num_Set
r.Df4 = r.Df2 - 1
r.TSP = dSumXY - dSumX * dSumY / r.GrandSum
r.ASP = r.ASP - dSumX * dSumY / r.GrandSum
r.WSP = r.TSP - r.ASP
r.AMSx = r.ASSx / r.Df1
r.AMSy = r.ASSy / r.Df1
r.WMSx = r.WSSx / r.Df2
```

Listing 9.1 (*Continued*)

```
  r.WMSy = r.WSSy / r.Df2
  r.Fx = r.AMSx / r.WMSx
  r.Fy = r.AMSy / r.WMSy
  r.TSSyhat = r.TSSy - r.TSP ^ 2 / r.TSSx
  r.WSSyhat = r.WSSy - r.WSP ^ 2 / r.WSSx
  r.ASSyhat = r.TSSyhat - r.WSSyhat
  r.AMSyhat = r.ASSyhat / r.Df3
  r.WMSyhat = r.WSSyhat / r.Df4
  r.ANOCOV_F = r.AMSyhat / r.WMSyhat
End Sub

Sub ANOVA1 (r As ANOVA1rec, DataMat() As Double, _
NData As Integer, NVar As Integer, hasMissingData As Integer, _
missingCode As Double, sum() As Double, sumX() As Double, _
sumX2() As Double)
Dim i As Integer, j As Integer
Dim x As Double
Dim TS As Double, TSS As Double
Dim GrandTotal As Double
Dim Mean() As Double
Dim Sdev() As Double

ReDim Mean(NVar)
ReDim Sdev(NVar)

For j = 0 To NVar - 1
  sum(j) = 0
  sumX(j) = 0
  sumX2(j) = 0
Next j
r.hasMissingData = hasMissingData
r.missingCode = missingCode

' update statistical summations
r.NVar = NVar
For i = 0 To NData - 1
  For j = 0 To NVar - 1
   x = DataMat(i, j)
   If Not (r.hasMissingData And x <= r.missingCode) Then
    sum(j) = sum(j) + 1
    sumX(j) = sumX(j) + x
    sumX2(j) = sumX2(j) + x * x
   End If
  Next j
Next i

' perform one way ANOVA calculations
TS = 0
TSS = 0
GrandTotal = 0
For i = 0 To r.NVar - 1
  Mean(i) = sumX(i) / sum(i)
  Sdev(i) = Sqr((sumX2(i) - sumX(i) ^ 2 / sum(i)) / _
      (sum(i) - 1))
  TS = TS + sumX(i)
  TSS = TSS + sumX2(i)
  GrandTotal = GrandTotal + sum(i)
Next i
r.Total_SS = TSS - TS ^ 2 / GrandTotal
r.Treatment_SS = 0
For i = 0 To r.NVar - 1
  r.Treatment_SS = r.Treatment_SS + sumX(i) ^ 2 / sum(i)
Next i
```

Listing 9.1   *(Continued)*

```
      r.Treatment_SS = r.Treatment_SS - TS ^ 2 / GrandTotal
      r.Error_SS = r.Total_SS - r.Treatment_SS
      r.Treatment_df = r.NVar - 1
      r.Error_df = GrandTotal - r.Treatment_df - 1
      r.Total_df = r.Error_df + r.Treatment_df
      r.Treatment_MS = r.Treatment_SS / r.Treatment_df
      r.Error_MS = r.Error_SS / r.Error_df
      r.ANOVA1_F = r.Treatment_MS / r.Error_MS
End Sub

Sub ANOVA2 (r As ANOVA2rec, DataMat() As Double, _
 NData As Integer, NVar As Integer, ColSumX() As Double, _
 ColSumX2() As Double, RowSumX() As Double, _
 RowSumX2() As Double)
Dim i As Integer, j As Integer
Dim nCol As Integer, nRow As Integer
Dim RowCol As Double
Dim x As Double, sx As Double

  ' initialize statistical summations
  For j = 0 To NVar - 1
    ColSumX(j) = 0
    ColSumX2(j) = 0
  Next j

  For j = 0 To NData - 1
    RowSumX(j) = 0
    RowSumX2(j) = 0
  Next j
  r.Num_Row = NData
  r.Num_Col = NVar

  ' update statistical summations
  r.GrandSumX = 0
  r.GrandSumX2 = 0
  r.Num_Col = NVar
  r.Num_Row = NData
  For i = 0 To NData - 1
    For j = 0 To NVar - 1
     x = DataMat(i, j)
     sx = x * x
     RowSumX(i) = RowSumX(i) + x
     ColSumX(j) = ColSumX(j) + x
     RowSumX2(i) = RowSumX2(i) + sx
     ColSumX2(j) = ColSumX2(j) + sx
    Next j
  Next i

  For i = 0 To NData - 1
    r.GrandSumX = r.GrandSumX + RowSumX(i)
    r.GrandSumX2 = r.GrandSumX2 + RowSumX2(i)
  Next i
  ' perform two-way ANOVA calculations
  nCol = r.Num_Col
  nRow = r.Num_Row
  RowCol = r.Num_Row * r.Num_Col
  r.Total_SS = r.GrandSumX2 - r.GrandSumX ^ 2 / RowCol
  r.Row_SS = 0
  For i = 0 To nRow - 1
    r.Row_SS = r.Row_SS + RowSumX(i) ^ 2 / nCol
  Next i
  r.Row_SS = r.Row_SS - r.GrandSumX ^ 2 / RowCol
  r.Col_SS = 0
```

Listing 9.1 (*Continued*)

```
  For i = 0 To nCol - 1
    r.Col_SS = r.Col_SS + ColSumX(i) ^ 2 / nRow
  Next i
  r.Col_SS = r.Col_SS - r.GrandSumX ^ 2 / RowCol
  r.Error_SS = r.Total_SS - r.Row_SS - r.Col_SS
  r.Row_df = r.Num_Row - 1
  r.Col_df = r.Num_Col - 1
  r.Error_df = r.Row_df * r.Col_df
  r.Total_df = RowCol - 1
  r.Row_F = (r.Row_SS / r.Row_df) / (r.Error_SS / r.Error_df)
  r.Col_F = (r.Col_SS / r.Col_df) / (r.Error_SS / r.Error_df)
End Sub

Sub ANOVA2R (r As ANOVA2Rrec, DataMat() As Double, _
  NData As Integer, NVar As Integer, NumReplicates As Integer, _
  ColSumX() As Double, RowSumX() As Double)
  Dim x As Double, sx As Double
  Dim nCol As Integer, nRow As Integer
  Dim RowCol As Double, sum As Double
  Dim i As Integer, j As Integer
  Dim k As Integer, iRow As Integer
  Dim CellSum() As Double

  ' initialize statistical summations
  ReDim CellSum(0 To NData, 0 To NVar)
  For j = 0 To NVar - 1
    ColSumX(j) = 0
  Next j
  For j = 0 To NData - 1
    RowSumX(j) = 0
  Next j
  For i = 0 To NData - 1
    For j = 0 To NVar - 1
      CellSum(i, j) = 0
    Next j
  Next i
  r.Num_Row = NData
  r.Num_Col = NVar
  r.GrandSumX = 0
  r.GrandSumX2 = 0

  ' update statistical summations
  r.NReplicate = NumReplicates
  For i = 0 To NData - 1
    iRow = i \ r.NReplicate
    For j = 0 To NVar - 1
      x = DataMat(i, j)
      sx = x * x
      CellSum(iRow, j) = CellSum(iRow, j) + x
      RowSumX(iRow) = RowSumX(iRow) + x
      ColSumX(j) = ColSumX(j) + x
      r.GrandSumX2 = r.GrandSumX2 + sx
    Next j
  Next i
  k = NData / r.NReplicate
  r.Num_Row = k
  r.Num_Col = NVar
  For j = 0 To NVar - 1
    r.GrandSumX = r.GrandSumX + ColSumX(j)
  Next j
  ' perform two-way ANOVA calculations
  nCol = r.Num_Col
  nRow = r.Num_Row
```

Listing 9.1 (*Continued*)

```
RowCol = r.Num_Row * r.Num_Col * r.NReplicate
r.Total_SS = r.GrandSumX2 - r.GrandSumX ^ 2 / RowCol
r.Row_SS = 0
For i = 0 To nRow - 1
  r.Row_SS = r.Row_SS + RowSumX(i) ^ 2 / _
      (r.Num_Col * r.NReplicate)
Next i
r.Row_SS = r.Row_SS - r.GrandSumX ^ 2 / RowCol
r.Col_SS = 0
For i = 0 To nCol - 1
  r.Col_SS = r.Col_SS + ColSumX(i) ^ 2 / _
      (r.Num_Row * r.NReplicate)
Next i
r.Col_SS = r.Col_SS - r.GrandSumX ^ 2 / RowCol
sum = 0
For i = 0 To nRow - 1
  For j = 0 To nCol - 1
    sum = sum + CellSum(i, j) ^ 2
  Next j
Next i
r.Interact_SS = sum / r.NReplicate - r.Row_SS - _
      r.Col_SS - r.GrandSumX ^ 2 / RowCol
r.Error_SS = r.GrandSumX2 - sum / r.NReplicate
r.Total_SS = r.GrandSumX2 - r.GrandSumX ^ 2 / RowCol
r.Row_df = r.Num_Row - 1
r.Col_df = r.Num_Col - 1
r.Interact_df = r.Row_df * r.Col_df
r.Total_df = RowCol - 1
r.Error_df = r.Num_Row * r.Num_Col * (r.NReplicate - 1)
r.Row_F = (r.Row_SS / r.Error_SS) * (r.Error_df / r.Row_df)
r.Col_F = (r.Col_SS / r.Error_SS) * (r.Error_df / r.Col_df)
r.Interact_F = (r.Interact_SS / r.Error_SS) * _
      (r.Error_df / r.Interact_df)
End Sub

Sub Latin (r As LatinSqrrec, DataMat() As Double, _
Map() As Integer, NData As Integer, ColSumX() As Double, _
RowSumX() As Double, TrtSumX() As Double)
Dim i As Integer, j As Integer, k As Integer
Dim x As Double, sx As Double, g As Double

' initialize statistical summations
For i = 0 To NData - 1
  ColSumX(i) = 0
  RowSumX(i) = 0
  TrtSumX(i) = 0
Next i
r.GrandSumX = 0
r.GrandSumX2 = 0
r.Num = 0

' update statistical summations
r.N = NData
' Obtain sums of columns, rows and treatments
For i = 0 To r.N - 1
  r.Num = r.Num + 1
  For j = 0 To r.N - 1
    x = DataMat(i, j)
    sx = x * x
    ColSumX(j) = ColSumX(j) + x
    RowSumX(i) = RowSumX(i) + x
    k = Map(i + NData * j)
    TrtSumX(k) = TrtSumX(k) + x
```

Listing 9.1 *(Continued)*

```
     r.GrandSumX2 = r.GrandSumX2 + sx
   Next j
   r.GrandSumX = r.GrandSumX + RowSumX(i)
 Next i

 ' perform Latin Squares ANOVA
 r.Row_SS = 0
 r.Col_SS = 0
 r.Trt_SS = 0
 For i = 0 To r.N - 1
   r.Row_SS = r.Row_SS + RowSumX(i) ^ 2
   r.Col_SS = r.Col_SS + ColSumX(i) ^ 2
   r.Trt_SS = r.Trt_SS + TrtSumX(i) ^ 2
 Next i
 g = (r.GrandSumX / r.Num) ^ 2
 r.Row_SS = r.Row_SS / r.Num - g
 r.Col_SS = r.Col_SS / r.Num - g
 r.Trt_SS = r.Trt_SS / r.Num - g
 r.Error_SS = r.GrandSumX2 - r.Row_SS - r.Col_SS - r.Trt_SS - g
 r.Total_SS = r.GrandSumX2 - g
 r.Row_df = r.Num - 1
 r.Col_df = r.Num - 1
 r.Trt_df = r.Num - 1
 r.Error_df = (r.Num - 1) * (r.Num - 2)
 r.Total_df = r.Num ^ 2 - 1
 r.Row_F = (r.Num - 2) * r.Row_SS / r.Error_SS
 r.Col_F = (r.Num - 2) * r.Col_SS / r.Error_SS
 r.Trt_F = (r.Num - 2) * r.Trt_SS / r.Error_SS
 End Sub
```

Listing 9.1 declares a set of user-defined structures that support the different ANOVA statistics. These structures are ANOVA1rec, ANOVA2rec, ANOVA2Rrec, LatinSqr, and ANOCOVrec. These identifiers support the one-way ANOVA, two-way ANOVA, two-way ANOVA with replication, Latin-Square ANOVA, and ANOCOV, respectively.

The module file declares the following Visual Basic functions:

1. The subroutine ANOVA1 supports the one-way ANOVA. This subroutine has parameters that specify an ANOVA1rec structure, the data matrix, the number of observations, the number of variables, the missing-code flag, the missing-code value, and the arrays of statistical summations.

2. The subroutine ANOVA2 supports the two-way ANOVA. This subroutine has parameters that specify an ANOVA2rec structure, the data matrix, the number of observations, the number of variables, and the arrays of statistical summations.

3. The subroutine ANOVA2R supports the two-way ANOVA with replication. This subroutine has parameters that specify an ANOVA2Rrec structure, the data matrix, the number of observations, the number of variables, the number of replicates, and the arrays of statistical summations.

4. The subroutine Latin supports the Latin-square ANOVA. This subroutine has parameters that specify a LatinSqr structure, the data matrix, the number of observations, the array that maps the indices, the number of replicates, and the arrays of statistical summations.

5. The subroutine ANOCOV supports the analysis of covariance. This subroutine has parameters that specify an ANOCOVrec structure, the data matrix, the number of observations, the number of sets, the missing-code flag, the missing-code value, and the arrays of statistical summations.

The Visual Basic Test Program

Let's look at the Visual Basic code that implements the optimization library. Listing 9.2 shows the source code for the form in project file TSANOVA.MAK. To compile the test program, you need to include the file ANOVA.BAS in your project file.

Listing 9.2 The source code for the form associated with the program project TSANOVA.MAK.

```
Const MISSING_DATA = -1E+30

Sub ANOCOVMnu_Click ()

  Static mat(10, 10) As Double
  Static sum(10) As Double
  Static sumX(10) As Double
  Static sumY(10) As Double
  Static sumX2(10) As Double
  Static sumY2(10) As Double
  Static sumXY(10) As Double
  Dim r5 As ANOCOVrec

  mat(0, 0) = 3
  mat(1, 0) = 2
  mat(2, 0) = 1
  mat(3, 0) = 2

  mat(0, 1) = 10
  mat(1, 1) = 8
  mat(2, 1) = 8
  mat(3, 1) = 11

  mat(0, 2) = 4
  mat(1, 2) = 3
  mat(2, 2) = 3
  mat(3, 2) = 5

  mat(0, 3) = 12
  mat(1, 3) = 12
  mat(2, 3) = 10
  mat(3, 3) = 13

  mat(0, 4) = 1
  mat(1, 4) = 2
  mat(2, 4) = 3
  mat(3, 4) = 1

  mat(0, 5) = 6
  mat(1, 5) = 5
  mat(2, 5) = 8
  mat(3, 5) = 7

  WindowState = 2
  Cls
  Print "************ ANOCOV **********"
  Print
  ANOCOV r5, mat(), 4, 3, False, 0, sum(), sumX(), sumY(), _
```

Listing 9.2 (*Continued*)

```
      sumX2(), sumY2(), sumXY()
 Print "Df among means = "; r5.Df1
 Print "Df within groups = "; r5.Df2
 Print "ASSx = "; r5.ASSx
 Print "ASP = "; r5.ASP
 Print "ASSy = "; r5.ASSy
 Print "WSSx = "; r5.WSSx
 Print "WSP = "; r5.WSP
 Print "WSSy = "; r5.WSSy
 Print "TSSx = "; r5.TSSx
 Print "TSP = "; r5.TSP
 Print "TSSy = "; r5.TSSy
 Print "------- Residuals statsitics ---------"
 Print "Df among means = "; r5.Df3
 Print "Df within groups = "; r5.Df4
 Print "ASSy^ = "; r5.ASSyhat
 Print "AMSy^ = "; r5.AMSyhat
 Print "WSSy^ = "; r5.WSSyhat
 Print "WMSy^ = "; r5.WMSyhat
 Print "TSSy^ = "; r5.TSSyhat
 Print "Fx = "; r5.Fy
 Print "Fy = "; r5.Fy
 Print "F = "; r5.ANOCOV_F
End Sub

Sub ANOVA1Mnu_Click ()
 Static mat(10, 10) As Double
 Static sum(10) As Double
 Static sumX(10) As Double
 Static sumX2(10) As Double
 Dim r1 As ANOVA1rec

 mat(0, 0) = 88
 mat(1, 0) = 99
 mat(2, 0) = 96
 mat(3, 0) = 68
 mat(4, 0) = 85
 mat(5, 0) = 2 * MISSING_DATA
 mat(6, 0) = 2 * MISSING_DATA

 mat(0, 1) = 78
 mat(1, 1) = 62
 mat(2, 1) = 98
 mat(3, 1) = 83
 mat(4, 1) = 61
 mat(5, 1) = 88
 mat(6, 1) = 2 * MISSING_DATA

 mat(0, 2) = 80
 mat(1, 2) = 61
 mat(2, 2) = 74
 mat(3, 2) = 92
 mat(4, 2) = 78
 mat(5, 2) = 54
 mat(6, 2) = 77

 mat(0, 3) = 71
 mat(1, 3) = 65
 mat(2, 3) = 90
 mat(3, 3) = 46
 mat(4, 3) = 2 * MISSING_DATA
 mat(5, 3) = 2 * MISSING_DATA
 mat(6, 3) = 2 * MISSING_DATA
```

Listing 9.2 (*Continued*)

```
ANOVA1 r1, mat(), 7, 4, True, MISSING_DATA, sum(), _
    sumX(), sumX2()
Cls
Print "************* one-way ANOVA *********"
Print
Print "Treatment SS = "; r1.Treatment_SS
Print "Treatment df = "; r1.Treatment_df
Print "Treatment MS = "; r1.Treatment_MS
Print "Error SS = "; r1.Error_SS
Print "Error df = "; r1.Error_df
Print "Error MS = "; r1.Error_MS
Print "Total SS = "; r1.Total_SS
Print "Total df = "; r1.Total_df
Print "F = "; r1.ANOVA1_F
End Sub

Sub ANOVA2Mnu_Click ()
 Static mat(10, 10) As Double
 Static ColSumX(10) As Double
 Static ColSumX2(10) As Double
 Static RowSumX(10) As Double
 Static RowSumX2(10) As Double
 Dim r2 As ANOVA2rec

 mat(0, 0) = 7
 mat(1, 0) = 2
 mat(2, 0) = 4
 mat(0, 1) = 6
 mat(1, 1) = 4
 mat(2, 1) = 6
 mat(0, 2) = 8
 mat(1, 2) = 4
 mat(2, 2) = 5
 mat(0, 3) = 7
 mat(1, 3) = 4
 mat(2, 3) = 3

 Cls
 Print "************* two-way ANOVA **********"
 Print
 ANOVA2 r2, mat(), 3, 4, ColSumX(), ColSumX2(), _
     RowSumX(), RowSumX2()
 Print "Row SS = "; r2.Row_SS
 Print "Row df = "; r2.Row_df
 Print "Row F = "; r2.Row_F
 Print "Col SS = "; r2.Col_SS
 Print "Col df = "; r2.Col_df
 Print "Col F = "; r2.Col_F
 Print "Error SS = "; r2.Error_SS
 Print "Error df = "; r2.Error_df
 Print "Total SS = "; r2.Total_SS
 Print "Total df = "; r2.Total_df
End Sub

Sub ANOVA2RMnu_Click ()
 Static mat(10, 10) As Double
 Static ColSumX(10) As Double
 Static RowSumX(10) As Double
 Dim r3 As ANOVA2Rrec

 mat(0, 0) = 2
 mat(1, 0) = 1.5
 mat(2, 0) = 1
```

Listing 9.2 (*Continued*)

```
mat(3, 0) = 1
mat(4, 0) = 1.5
mat(5, 0) = 1

mat(0, 1) = 1
mat(1, 1) = 1.5
mat(2, 1) = 0
mat(3, 1) = 1
mat(4, 1) = 1
mat(5, 1) = 1.5

mat(0, 2) = -.5
mat(1, 2) = .5
mat(2, 2) = -1
mat(3, 2) = 0
mat(4, 2) = 1
mat(5, 2) = 1

mat(0, 3) = 1.5
mat(1, 3) = 1.5
mat(2, 3) = -1
mat(3, 3) = 0
mat(4, 3) = .5
mat(5, 3) = 1

Cls
Print "*** two-way ANOVA with replication ***"
Print
ANOVA2R r3, mat(), 6, 4, 2, ColSumX(), RowSumX()
Print "Row SS = "; r3.Row_SS
Print "Row df = "; r3.Row_df
Print "Row F = "; r3.Row_F
Print "Col SS = "; r3.Col_SS
Print "Col df = "; r3.Col_df
Print "Col F = "; r3.Col_F
Print "Interaction SS = "; r3.Interact_SS
Print "Interaction df = "; r3.Interact_df
Print "Interaction F = "; r3.Interact_F
Print "Error SS = "; r3.Error_SS
Print "Error df = "; r3.Error_df
Print "Total SS = "; r3.Total_SS
Print "Total df = "; r3.Total_df
End Sub

Sub ExitMnu_Click ()
 End
End Sub

Sub Form_Load ()
 AutoRedraw = True
End Sub

Sub LatinMnu_Click ()
 Static mat(10, 10) As Double
 Static map(25) As Integer
 Static ColSumX(10) As Double
 Static RowSumX(10) As Double
 Static TrtSumX(10) As Double
 Dim r4 As LatinSqrrec

 map(0) = 0
 map(1) = 2
 map(2) = 1
```

Listing 1.1 *(Continued)*

```
map(3)  = 3
map(4)  = 1
map(5)  = 3
map(6)  = 0
map(7)  = 2
map(8)  = 2
map(9)  = 0
map(10) = 3
map(11) = 1
map(12) = 3
map(13) = 1
map(14) = 2
map(15) = 0

mat(0, 0) = 26.46
mat(0, 1) = 29.61
mat(0, 2) = 27.82
mat(0, 3) = 29.15
mat(1, 0) = 27.58
mat(1, 1) = 29.52
mat(1, 2) = 26.48
mat(1, 3) = 29.13
mat(2, 0) = 29.54
mat(2, 1) = 27#
mat(2, 2) = 29.31
mat(2, 3) = 27.9
mat(3, 0) = 29.15
mat(3, 1) = 28.03
mat(3, 2) = 29.53
mat(3, 3) = 26.51

Cls
Print "*********** Latin Square ANOVA *************"
Print
Latin r4, mat(), map(), 4, ColSumX(), RowSumX(), TrtSumX()
Print "Row SS = "; r4.Row_SS
Print "Row df = "; r4.Row_df
Print "Row F = "; r4.Row_F
Print "Col SS = "; r4.Col_SS
Print "Col df = "; r4.Col_df
Print "Col F = "; r4.Col_F
Print "Treatment SS = "; r4.Trt_SS
Print "Treatment df = "; r4.Trt_df
Print "Treatment F = "; r4.Trt_F
Print "Error SS = "; r4.Error_SS
Print "Error df = "; r4.Error_df
Print "Total SS = "; r4.Total_SS
Print "Total df = "; r4.Total_df
End Sub
```

The project uses a form that has a simple menu system but no controls. Table 9.1 shows the menu structure and the names of the menu items. The form has the caption "Analysis of Variance." The menu option Test has five selections to test the various ANOVA tests. Each one of these menu selections clears the form and then displays the result of the tested ANOVA subroutines. Thus, you can zoom in on any method by invoking its related menu selection.

TABLE 9.1 The Menu System for the TSANOVA.MAK Project

Menu caption	Name
&Test	TesMnu
One-Way ANOVA	ANOVA1Mnu
Two-Way ANOVA (no replicates)	ANOVA2Mnu
Two-Way ANOVA (with replicates)	ANOVA2RMnu
Latin Square ANOVA	LatinMnu
ANOCOV	ANOCOVMnu
–	N1
&Exit	ExitMnu

The program tests the various ANOVA-related subroutines using matrices of data. The program assigns different data for testing each Visual Basic function. Figure 9.1 shows the output of the test program for each menu selection.

```
************* one-way ANOVA *********

Treatment SS = 930.438095238118
Treatment df = 3
Treatment MS = 310.146031746039
Error SS = 3599.56190476188
Error df = 18
Error MS = 199.97566137566
Total SS = 4530
Total df = 21
F = 1.55091889489202


************* two-way ANOVA **********

Row SS = 26
Row df = 2
Row F = 11.7
Col SS = 3.33333333333331
Col df = 3
Col F = .999999999999992
Error SS = 6.66666666666669
Error df = 6
Total SS = 36
Total df = 11


*** two-way ANOVA with replication ***

Row SS = 5.02083333333333
Row df = 2
Row F = 11.4761904761905
```

Figure 9.1 The output of the sample program for ANOVA tests.

```
Col SS = 4.61458333333333
Col df = 3
Col F = 7.03174603174603
Interaction SS = 2.72916666666667
Interaction df = 6
Interaction F = 2.07936507936508
Error SS = 2.625
Error df = 12
Total SS = 14.9895833333333
Total df = 23


*********** Latin Square ANOVA *************

Row SS = .141749999997046
Row df = 3
Row F = 5.58070866120319
Col SS = .351149999998597
Col df = 3
Col F = 13.8248031493087
Treatment SS = 21.4386999999988
Treatment df = 3
Treatment F = 844.04330707177
Error SS = 5.08000000008906E-02
Error df = 6
Total SS = 21.9823999999953
Total df = 15


************* ANOCOV **********

Df among means = 2
Df within groups = 9
ASSx = 9.5
ASP = 20.75
ASSy = 55.1666666666667
WSSx = 7.5
WSP = 6.25
WSSy = 16.5
TSSx = 17
TSP = 27
TSSy = 71.6666666666667
------- Residuals statistics ---------
Df among means = 2
Df within groups = 8
ASSy^ = 17.4926470588235
AMSy^ = 8.74632352941176
WSSy^ = 11.2916666666667
WMSy^ = 1.41145833333333
TSSy^ = 28.7843137254902
Fx = 15.0454545454545
Fy = 15.0454545454545
F = 6.19665726069025
```

Figure 9.1 (*Continued*)

Chapter 10

Linear Regression

Linear regression enables you to examine the relationship between two variables. The basic linear regression calculations yield the best line that passes through the observations. By applying mathematical transformations to the variables, you can also obtain the best fit for different kinds of curves. This chapter looks at the following aspects of linear regression:

- Basic linear regression
- Linearized regression
- The confidence interval for the project values
- The confidence interval for the regression coefficients
- Testing the values of the regression coefficients
- The automatic best method

Basic Linear Regression

The basic linear regression examines the correlation between an independent variable, x, and a dependent variable y. Linear regression attempts to find the slope and intercept for the best line that passes through the observed values of x and y. This simple relation is given by the following equation:

$$y = A + B \, x \qquad (10.1)$$

where A is the intercept and B is the slope. Figure 10.1 shows a sample linear regression.

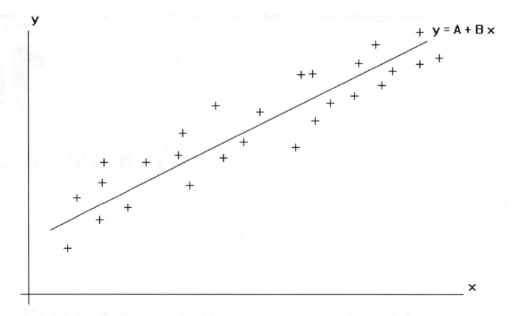

Figure 10.1 A sample linear regression.

To calculate the slope for the best line, use the following equation:

$$B = \frac{(n\Sigma x_i y_i - (\Sigma x_i)(\Sigma y_i))}{(n\Sigma x_i^2 - (\Sigma x_i)^2)}$$ (10.2)

The following equation calculates the intercept A:

$$A = \frac{(\Sigma y_i - B\,\Sigma x_i)}{n}$$ (10.3)

The coefficient of determination, R^2, indicates the percent (as a fraction) of the variation between variables x and y, which is explained by equation 10.1. When R^2 is one, you have a perfect fit, and equation 10.1 fully explains the relation between variable x and y. By contrast, when R^2 is zero, there is no relation between the two variables.

The following equation shows how to calculate the value for R^2:

$$R^2 = \frac{[A\Sigma y_i + B\Sigma x_i y_i - (\Sigma y_i)^2/n]}{[\Sigma y_i^2 - (\Sigma y_i)^2/n]}$$ (10.4)

The ANOVA associated with linear regression calculates the following items:

- The variances:

$$S(X^2) = \Sigma x_i^2 - n(\Sigma x_i)^2$$ (10.5)
$$S(Y^2) = \Sigma y_i^2 - n(\Sigma y_i)^2$$ (10.6)
$$S(XY) = \Sigma x_i y_i - n(\Sigma x_i)(\Sigma y_i)$$ (10.7)

- The sum of squares for y, due to A + Bm$_x$, due to B, and the residuals where m$_x$ is the mean value for x:

$$SSY = \Sigma y_i^2 \tag{10.8}$$

$$SSA = \frac{(\Sigma y_i)^2}{n} \tag{10.9}$$

$$SSB = b\ S(XY) \tag{10.10}$$

$$SSD = S(Y2) - b\ S(XY) \tag{10.11}$$

- The degrees of freedom for Y, due to A + Bm$_x$, due to B, and the residuals:

$$dfY = n \tag{10.12}$$

$$dfA = 1 \tag{10.13}$$

$$dfB = 1 \tag{10.14}$$

$$ResDf = n - 2 \tag{10.15}$$

- The mean squares for the slope B and for the residuals:

$$MSB = b\ S(XY) \tag{10.16}$$

$$s_{ey}^2 = \frac{SSD}{(n-2)} \tag{10.17}$$

Linearized Regression

Equation 10.1 represents the simplest relationship between two variables. Often, you already know or strongly suspect that a pair of variables are related in a nonlinear fashion that can be linearized by applying the appropriate transformations. When this nonlinear relation can be linearized, you can use the previous equations on the transformed data. Thus, for example, if you have variables x and y related in the following power function:

$$y = A\ x^B \tag{10.18}$$

Then, by applying the ln function to both sides of equation 10.5, you can linearize it:

$$\ln(y) = \ln(A) + B \ln(x) \tag{10.19}$$

You can regard equation 10.19 as:

$$y' = A' + B\ x' \tag{10.20}$$

where y' is ln(y), A' is ln(A), and x' is ln(x). Thus, applying linear regression to ln(y) and ln(x) yields the intercept ln(A) and the slope B. You can also apply this method with other functions, such as the square, square root, and reciprocal, to obtain a linear fit.

The Confidence Interval for Projections

Once you calculate the slope and intercept for the best line, you can use equation 10.1 to calculated the projected values of y for given values of x (and vice versa).

What about the confidence interval for the project value of y? Here is the equation or calculating the confidence interval for y:

$$y_0 \pm t_{n-2;\alpha/2} \sqrt{\left[s_{ey}^2\left(\frac{1}{n} + \frac{(x_0 - m_x)^2}{S(X^2)}\right)\right]} \qquad (10.21)$$

where y_0 is the projected value of y, $t_{n-2;\alpha/2}$ is the Student-t value for $n-2$ degrees of freedom and α is the the risk probability ($1 - \alpha$ is the confidence probability). Figure 10.2 depicts a sample confidence interval band for the projected values of the dependent variable.

The Confidence Interval for Regression Coefficients

The slope and intercept values represent estimates for the mean value of the real slope and intercept. You can calculate the confidence interval for the slope and intercept using the following equations:

$$A \pm t_{n-2;\alpha/2} \sqrt{\left[s_{ey}^2\left(\frac{1}{n} + \frac{m_x^2}{S(X^2)}\right)\right]} \qquad (10.22)$$

$$B \pm t_{n-2;\alpha/2} \sqrt{\left[\frac{s_{ey}^2}{S(X^2)}\right]} \qquad (10.23)$$

You can also use equations 10.22 and 10.23 to test whether the slope and intercept statistically differ from specific values. As for the correlation coefficient (which is

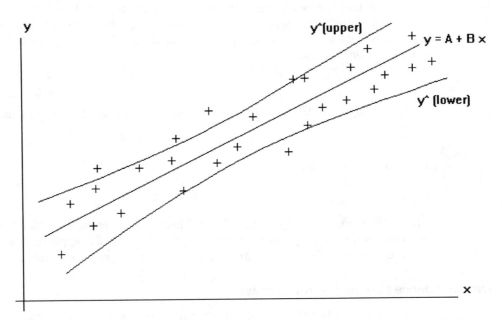

Figure 10.2 A sample depiction of the confidence interval band for the projected values of the dependent variable.

merely the square root of the determination coefficient), you can use the following inequality to test if the value of the correlation coefficient differs from zero:

$$|R| \frac{\sqrt{(n-2)}}{\sqrt{(1-R^2)}} > t_{n-2;\alpha/2} \qquad (10.24)$$

The Regression Algorithms

Let's look at the algorithms involved in linear (and linearized) regression. This process includes initializing the statistical summations, accumulating data in these summations, and calculating the regression coefficients and regression ANOVA statistics. Here is the algorithm for performing the linear regression:

Given:

- The matrix Xmat with N observations
- The indices Xindex and Yindex, which select the X and Y variables from matrix Xmat
- The transformation functions fx, fy, and invfy

Algorithm:

1. Set Sum, SumX, SumX2, SumY, SumY2, and SumXY to 0
2. For I = 0 to N − 1, repeat the next steps:
    2.1. Set Xr = Xmat[I, Xindex]
    2.2. Set Yr = Xmat[I, Yindex]
    2.3. If Xr or Yr represents missing data, resume at step 2
    2.4. Set Xr = fx(Xr)
    2.5. Set Yr = fy(Yr)
    2.6. Increment Sum
    2.7. Add Xr * Yr to SumXY
    2.8. Add Xr to SumX
    2.9. Add Yr to SumY
    2.10. Add Xr squared to SumX2
    2.11. Add Yr squared to SumY2

3. Set MeanX = $\dfrac{\text{SumX}}{\text{Sum}}$

4. Set MeanY = $\dfrac{\text{SumY}}{\text{Sum}}$

5. Set SdevX = $\sqrt{\left(\dfrac{(\text{SumXX} - \text{SumX}^2/\text{Sum})}{(\text{Sum} - 1)} \right)}$

6. Set SdevY = $\sqrt{\left(\dfrac{(\text{SumYY} - \text{SumY}^2/\text{Sum})}{(\text{Sum} - 1)} \right)}$

7. Set Slope = $\dfrac{(\text{SumXY} - \text{MeanX} * \text{MeanY} * \text{Sum})/\text{SdevX}^2}{(\text{Sum} - 1)}$

8. Set Intercept = MeanY – Slope * MeanX

9. Set R2 = $\left(\dfrac{\text{Slope} * \text{SdevX}}{\text{SdevY}}\right)^\wedge 2$

10. Set RegSS = $\dfrac{(\text{SumXY} - \text{SumY} * \text{MeanX})^\wedge 2}{(\text{SdevX}^\wedge 2 * (\text{Sum} - 1))}$

11. Set TotalSS = $\dfrac{\text{SumYY} - \text{SumY}^\wedge 2}{\text{Sum}}$

12. Set ResidualSS = TotalSS – RegSS

13. Set Residualdf = Sum – 2

14. Set S = $\sqrt{\left(\dfrac{\text{ResidualSS}}{\text{Residualdf}}\right)}$

15. Set StdErSlope = S / $\sqrt{\left(\dfrac{\text{SumXX} - \text{SumX}^\wedge 2}{\text{Sum}}\right)}$

16. Set StdErIntercept = S * $\sqrt{\left(\dfrac{\text{SumXX/Sum/SdevX}^\wedge 2}{(\text{Sum} - 1)}\right)}$

17. Set Regdf = 1

18. Set Totaldf = Sum – 1

19. Set S2 = S * S

20. F = $\dfrac{\text{RegSS}}{\text{S2}}$

The Automatic Best Fit

Computers are ideal for systematic number-crunching, so you can use them to systematically fit a set of data with various kinds of curves. Each curve requires a distinct combination of transformation functions. Moreover, each curve yields a set of regression coefficients and determination coefficient. The best curve-fit is the one that produces the highest value for the determination coefficient.

This chapter presents the source code for the automatic best fit that sorts the results of four curve-fits in order of descending determination coefficient values. If you apply square root, logarithm, and reciprocal functions, you need to make sure that the examined data contain only positive values.

The Visual Basic Source Code

Let's look at the Visual Basic source code that implements the regression analysis and related tests and calculations. Listing 10.1 shows the source code for the LIN-REG.BAS module file.

 Throughout the book, the underscore character is used to split wrapping lines of Visual Basic declarations and statements.

Listing 10.1 The source code for the LINREG.BAS module file.

```
' ANOVA table record
Type LinRegANOVA
 Reg_df As Double
 Reg_SS As Double
 Residual_df As Double
 Residual_SS As Double
 Total_df As Double
 Total_SS As Double
 S2 As Double
 LINREG_F As Double
End Type

' Linear regression record
Type LinRegrec
 Xindex As Integer
 Yindex As Integer
 hasMissingData As Integer
 missingCode As Double
 sumX As Double
 sumXX As Double
 sum As Double
 sumY As Double
 sumYY As Double
 sumXY As Double
 ' basic statistics
 MeanX As Double
 MeanY As Double
 SdevX As Double
 SdevY As Double
 Slope As Double
 Intercept As Double
 R2 As Double
 ' Regression results
 StdErSlope As Double
 StdErIntercept As Double
 ANOVA As LinRegANOVA
End Type

Type BestLRrec
 fxIndex As Integer
 R2 As Double
 Slope As Double
 Intercept As Double
End Type

Function BestFit (DataMat() As Double, NData As Integer, _
 numTrnsFx As Integer, Xindex As Integer, Yindex As Integer, _
 r() As BestLRrec) As Integer
' find the best model to fit the data

 Dim sum As Double, sumX As Double
 Dim sumXX As Double, sumY As Double
 Dim sumYY As Double, sumXY As Double
 Dim MeanX As Double, MeanY As Double
 Dim SdevX As Double, SdevY As Double
 Dim x As Double, y As Double
 Dim i As Integer, j As Integer
 Dim offset As Integer, inOrder As Integer
 Dim temp As BestLRrec

 If (Xindex = Yindex) Or (NData < 3) Or (numTrnsFx < 2) Then
   BestFit = False
   Exit Function
 End If
```

Listing 10.1 *(Continued)*

```
' iterate over the functions
For i = 0 To numTrnsFx - 1
  ' initialize summations
  sum = 0
  sumX = 0
  sumXX = 0
  sumY = 0
  sumYY = 0
  sumXY = 0
  ' process the observations
  For j = 0 To NData - 1
   x = DataMat(j, Xindex)
   y = DataMat(j, Yindex)
   x = MyBestLRFx(x, i)
   y = MyBestLRFy(y, i)
   sum = sum + 1
   sumX = sumX + x
   sumY = sumY + y
   sumXX = sumXX + x * x
   sumYY = sumYY + y * y
   sumXY = sumXY + x * y
  Next j
  ' calculate the results
  MeanX = sumX / sum
  MeanY = sumY / sum
  SdevX = Sqr((sumXX - sumX ^ 2 / sum) / (sum - 1))
  SdevY = Sqr((sumYY - sumY ^ 2 / sum) / (sum - 1))
  r(i).Slope = (sumXY - MeanX * MeanY * sum) / SdevX ^ 2 / _
        (sum - 1)
  r(i).Intercept = MeanY - r(i).Slope * MeanX
  r(i).R2 = (r(i).Slope * SdevX / SdevY) ^ 2
  r(i).fxIndex = i
Next i
' sort the results in descending order
offset = numTrnsFx
Do
  offset = (offset * 8) \ 11
  If offset = 0 Then offset = 1
  inOrder = True
  For i = 0 To numTrnsFx - offset - 1
   j = i + offset
   ' swap elements?
   If r(i).R2 < r(j).R2 Then
    inOrder = False
    temp = r(i)
    r(i) = r(j)
    r(j) = temp
   End If
  Next i
 Loop Until (offset = 1) And inOrder
 BestFit = True
End Function

Sub InitializeLinReg (r As LinRegrec, Xindex As Integer, _
 Yindex As Integer, hasMissingData As Integer, _
 missingCode As Double)
' initialize statistical summations and data range

 r.sumX = 0
 r.sumXX = 0
 r.sum = 0
 r.sumY = 0
 r.sumYY = 0
```

Listing 10.1 *(Continued)*

```
  r.sumXY = 0
  r.hasMissingData = hasMissingData
  r.missingCode = missingCode
  r.Xindex = Xindex
  r.Yindex = Yindex
End Sub

Sub LinReg (r As LinRegrec)
' calculate regression coefficients and related results

  Dim S As Double

  r.MeanX = r.sumX / r.sum
  r.MeanY = r.sumY / r.sum
  r.SdevX = Sqr((r.sumXX - r.sumX ^ 2 / r.sum) / (r.sum - 1))
  r.SdevY = Sqr((r.sumYY - r.sumY ^ 2 / r.sum) / (r.sum - 1))
  r.Slope = (r.sumXY - r.MeanX * r.MeanY * r.sum) / _
        r.SdevX ^ 2 / (r.sum - 1)
  r.Intercept = r.MeanY - r.Slope * r.MeanX
  r.R2 = (r.Slope * r.SdevX / r.SdevY) ^ 2
  r.ANOVA.Reg_SS = (r.sumXY - r.sumY * r.MeanX) ^ 2 / _
          ((r.SdevX) ^ 2 * (r.sum - 1))
  r.ANOVA.Total_SS = r.sumYY - r.sumY ^ 2 / r.sum
  r.ANOVA.Residual_SS = r.ANOVA.Total_SS - r.ANOVA.Reg_SS
  r.ANOVA.Residual_df = r.sum - 2
  S = Sqr(r.ANOVA.Residual_SS / r.ANOVA.Residual_df)
  r.StdErSlope = S / Sqr(r.sumXX - r.sumX ^ 2 / r.sum)
  r.StdErIntercept = S * Sqr(r.sumXX / r.sum / r.SdevX ^ 2 / _
            (r.sum - 1))
  r.ANOVA.Reg_df = 1
  r.ANOVA.Total_df = r.sum - 1
  r.ANOVA.S2 = S * S
  r.ANOVA.LINREG_F = r.ANOVA.Reg_SS / r.ANOVA.S2
End Sub

Sub LinRegCoefCI (r As LinRegrec, probability As Double, _
  slopeHi As Double, slopeLow As Double, intHi As Double, _
  intLow As Double)
' calculate confidence interval for slope and intercept

  Dim Df As Double, tableT As Double
  Dim diff As Double, p As Double

  If probability > 1 Then
    p = .5 - probability / 200
  Else
    p = .5 - probability / 2
  End If
  Df = r.sum - 2
  tableT = TInv(p, Df)
  diff = tableT * r.StdErSlope
  slopeHi = r.Slope + diff
  slopeLow = r.Slope - diff

  diff = tableT * r.StdErIntercept
  intHi = r.Intercept + diff
  intLow = r.Intercept - diff
End Sub

Sub LR_Int_T_test (r As LinRegrec, probability As Double, _
  testValue As Double, calcT As Double, tableT As Double, _
  passTest As Integer)
' compare intercept value with a tested value
```

Listing 10.1 *(Continued)*

```
'  Hypothesis tested is H0: Intercept = testValue

 Dim Df As Double, p As Double

 If probability > 1 Then
   p = .5 - probability / 200
 Else
   p = .5 - probability / 2
 End If
 Df = r.sum - 2
 tableT = TInv(Df, p)
 calcT = (r.Intercept - testValue) / r.StdErIntercept
 passTest = Abs(calcT) <= tableT
End Sub

Sub LR_R2_T_Test (r As LinRegrec, probability As Double, _
 calcT As Double, tableT As Double, passTest As Integer)
' Procedure to test hypothesis H0 : R^2 = 0

 Dim Df As Double, p As Double

 If probability > 1 Then
   p = .5 - probability / 200
 Else
   p = .5 - probability / 2
 End If
 Df = r.sum - 2
 tableT = TInv(p, Df)
 calcT = Sqr(r.R2 * Df / (1 - r.R2))
 passTest = calcT <= tableT
End Sub

Sub LR_Slope_T_test (r As LinRegrec, probability As Double, _
 testValue As Double, calcT As Double, tableT As Double, _
 passTest As Integer)
' compare slope value with a tested value
'  Hypothesis tested is H0: Slope = testValue

 Dim Df As Double, p As Double

 If probability > 1 Then
   p = .5 - probability / 200
 Else
   p = .5 - probability / 2
 End If
 Df = r.sum - 2
 tableT = TInv(p, Df)
 calcT = (r.Slope - testValue) / r.StdErSlope
 passTest = Abs(calcT) <= tableT
End Sub

Sub sumLinReg (DataMat() As Double, r As LinRegrec, _
 NData As Integer)
' update statistical summations

 Dim i As Integer
 Dim xr As Double, yr As Double

 For i = 0 To NData - 1
   xr = DataMat(i, r.Xindex)
   yr = DataMat(i, r.Yindex)
   If Not (r.hasMissingData And ((xr <= r.missingCode) Or _
     (yr <= r.missingCode))) Then
```

Listing 10.1 (*Continued*)

```
    ' transform x and y data
    xr = MyLRFx(xr)
    yr = MyLRFy(yr)
    ' Update summations
    r.sum = r.sum + 1
    r.sumXY = r.sumXY + xr * yr
    r.sumX = r.sumX + xr
    r.sumXX = r.sumXX + xr ^ 2
    r.sumY = r.sumY + yr
    r.sumYY = r.sumYY + yr ^ 2
  End If
 Next i
End Sub

Sub YhatCI (r As LinRegrec, xr As Double, _
 probability As Double, yHat As Double, yHi As Double, _
 yLow As Double)
' calculate projections and their confidence interval

 Dim Df As Double, deltaY As Double
 Dim p As Double, tableT As Double

 If probability > 1 Then
   p = .5 - probability / 200
 Else
   p = .5 - probability / 2
 End If
 Df = r.sum - 2
 tableT = TInv(p, Df)
 xr = MyLRFx(xr) ' transform xr
 deltaY = Sqr((xr - r.MeanX) ^ 2 / (r.sumX ^ 2 * (r.sum - 1)) _
      + 1 / (r.sum + 1)) * Sqr(r.ANOVA.S2) * tableT
 yHat = r.Intercept + r.Slope * xr
 yHi = MyLRInvFy(yHat + deltaY)
 yLow = MyLRInvFy(yHat - deltaY)
 yHat = MyLRInvFy(yHat)
End Sub
```

Listing 10.1 shows the declaration of data types LinRegANOVA, LinRegrec, and BestLinReg. The data type LinRegANOVA stores the fields for the regression ANOVA. The data type LinRegrec stores the fields for the linearized regression. The data type BestLinReg stores the data for the current fit. The listing declares the following Visual Basic functions:

1. The subroutine InitializeLinReg initializes the LinRegrec-type parameter r. This subroutine includes additional parameters that specify the indices for the regression variables, the missing-code flag, and the missing-code value. You must call this subroutine at least once to initialize the LinRegrec-type variable used in the regression. Additional calls to this subroutine (with the same LinRegrec-type variable) help you reset the targeted regression structure and prepare it for another round of calculations.

2. The subroutine sumLinReg updates the statistical summations. This subroutine has parameters that specify the data matrix, the regression's LinRegrec-type parameter, and the number of observations. You must call this subroutine with enough data at least once. You can call this subroutine multiple times to accumu-

late a large number of observations. Keep in mind that in this case, the data matrix should pass data with the same number of variables.

3. The subroutine LinReg performs the linear regression calculations. This subroutine has a single LinRegrec-type parameter that passes data to and from the function LinReg. You must call this subroutine before calling the next five subroutines, since these subroutines expect the accessed LinRegrec-type parameter to contain regression results. The subroutine LinReg uses the user-defined functions MyFx and MyFy (defined in module MYLINREG.BAS) to perform the transformations on the data.

4. The subroutine YhatCI calculates the projected y value and its confidence interval. This subroutine has parameters that access a LinRegrec-type variable, the value of x to project, the probability for the confidence interval, the projected y, the lower interval value of y, and the higher interval value of y. The subroutine YhatCI uses the user-defined functions MyFx, MyFy, and MyInvFy (defined in module MYLINREG.BAS) to perform the transformations on the data.

5. The subroutine LinRegCoefCI calculates the confidence intervals for the regression slope and intercept. The subroutine has parameters that access a LinRegrec-type variable, the probability for the confidence intervals, and the low and high limits for each of the slope and intercept.

6. The subroutine LR_Slope_T_test tests the value of the slope. This subroutine has parameters that access a LinRegrec-type variable, the probability for the confidence intervals, the tested slope value, the calculated Student-t value, the tabulated Student-t value, and a Boolean flag. The latter flag indicates whether or not the tested value is accepted.

7. The subroutine LR_Int_T_test tests the value of the intercept. This subroutine has parameters that access a LinRegrec-type variable, the probability for the confidence intervals, the tested intercept value, the calculated Student-t value, the tabulated Student-t value, and a Boolean flag. The latter flag indicates whether or not the tested value is accepted.

8. The subroutine LR_R2_T_test tests whether the value of R (or R^2) is zero. The subroutine has parameters that access a LinRegrec-type variable, the probability for the test, the calculated Student-t value, the tabulated Student-t value, and a Boolean flag.

9. The function BestFit performs the automatic best fit. The function has parameters that specify the data matrix, number of observations, the number of transformation functions, the index of variable x, the index of variable y, the array of transformation functions for x, the array of transformation functions for y, and the array of BestLRrec data type. The latter array passes the results of the automatic best fit to the caller. The function sorts these results in the order of descending values of R^2. The function BestFit uses the user-defined functions MyBestLRFx and MyBestLRFy (defined in module MYLINREG.BAS) to select the transformations.

Listing 10.2 shows the source code for the MYLINREG.BAS module file. This file contains the user-defined functions that perform various transformations for the regression data.

Listing 10.2 The source code for the MYLINREG.BAS module file.

```
Function MyBestLRFx (X As Double, Index As Integer) As Double

  Select Case Index

   Case 0, 1
    MyBestLRFx = X

   Case 2, 3
    MyBestLRFx = Log(X)

   Case Else
    MyBestLRFx = X
  End Select
End Function

Function MyBestLRFy (Y As Double, Index As Integer) As Double

  Select Case Index

   Case 0, 2
    MyBestLRFy = Y

   Case 1, 3
    MyBestLRFy = Log(Y)

   Case Else
    MyBestLRFy = Y

  End Select
End Function

Function MyLRFx (X As Double) As Double
 MyLRFx = X
End Function

Function MyLRFy (Y As Double) As Double
 MyLRFy = Y
End Function

Function MyLRInvFy (Y As Double) As Double
 MyLRInvFy = Y
End Function
```

The function MyFx and MyFy in Listing 10.2 tranform the X and Y data, respectively. The function MyInvFy performs the reverse transformation of function MyFy. The current forms of these three Visual Basic functions do not transform the regression data.

The functions MyBestLRFx and MyBestLRFy perform the transformations for the X and Y data when performing the automatic best fit. Notice that these two functions use the Select Case statement with multiple-value Case labels. This approach enables you to perform the needed transformations without using arrays of indices that select the transformations. The current form of the functions MyBestLRFx and MyBest-LRFy support X versus Y, X versus ln(Y), ln(X) versus Y, and ln(X) versus ln(Y).

The Visual Basic Test Program

Listing 10.3 shows the source code for the form associated with the TSLINREG.MAK project. To compile the test program, you need to include the files LINREG.BAS, MYLINREG.BAS, and STATLIB.BAS in your project file.

Listing 10.3 The source code for the form associated with the TSLINREG.MAK project.

```
Const MAX_FX% = 4

Dim r As LinRegrec

Sub BestLRMnu_Click ()
 Static mat(4, 1) As Double
 Static bestR2(MAX_FX) As BestLRrec
 Dim i As Integer
 Dim numFx As Integer

 mat(0, 0) = 10
 mat(0, 1) = 2.3026
 mat(1, 0) = 25
 mat(1, 1) = 3.2189
 mat(2, 0) = 30
 mat(2, 1) = 3.4012
 mat(3, 0) = 35
 mat(3, 1) = 3.5553
 mat(4, 0) = 100
 mat(4, 1) = 4.6052
 numFx = MAX_FX

 dummy = BestFit(mat(), 5, numFx, 0, 1, bestR2())
 Cls
 Print "************** Automatic Best Fit ****************"
 Print
 Print "Function #";
 Print Tab(15); "R2";
 Print Tab(30); "Slope";
 Print Tab(45); "Intercept"
 Print
 For i = 0 To numFx - 1
  Print bestR2(i).fxIndex;
  Print Tab(15); Format$(bestR2(i).R2, "0.#####");
  Print Tab(30); Format$(bestR2(i).Slope, "#.####E+00");
  Print Tab(45); Format$(bestR2(i).Intercept, "#.####E+00")
 Next i
End Sub

Sub CIMnu_Click ()
 Dim probability As Double
 Dim slopeHi As Double
 Dim slopeLow As Double
 Dim intHi As Double
 Dim intLow As Double

 probability = 95
 LinRegCoefCI r, probability, slopeHi, slopeLow, intHi, intLow
 Cls
 Print "************* Confidence Interval *************"
 Print
 Print "At "; probability; "% probability"
 Print "Range for slope is "; slopeLow; "to "; slopeHi
 Print "Range for intercept is "; intLow; " to "; intHi
End Sub
```

Listing 10.3 (*Continued*)

```
Sub ExitMnu_Click ()
 End
End Sub

Sub Form_Load ()
 CIMnu.Enabled = False
 TestSlopeMnu.Enabled = False
 TestInterceptMnu.Enabled = False
 TestR2Mnu.Enabled = False
End Sub

Sub LRMnu_Click ()
 Static mat(6, 1) As Double
 Dim i As Integer

 mat(0, 0) = 2
 mat(0, 1) = 24.5
 mat(1, 0) = 4
 mat(1, 1) = 19.5
 mat(2, 0) = 6
 mat(2, 1) = 17.5
 mat(3, 0) = 8
 mat(3, 1) = 14
 mat(4, 0) = 10
 mat(4, 1) = 12
 mat(5, 0) = 12
 mat(5, 1) = 7
 mat(6, 0) = 14
 mat(6, 1) = 5

 InitializeLinReg r, 0, 1, False, 0
 sumLinReg mat(), r, 7
 LinReg r
 Cls
 Print "************* Linear Regression **********"
 Print
 Print "Number of points: "; r.sum
 Print "R^2 = "; Format$(r.R2, "0.#####")
 Print "Slope = "; r.Slope
 Print "Intercept = "; r.Intercept
 Print
 Print "************* ANOVA **********"
 Print "Regression SS = "; r.ANOVA.Reg_SS
 Print "Regression df = "; r.ANOVA.Reg_df
 Print "Residual SS = "; r.ANOVA.Residual_SS
 Print "Residual df = "; r.ANOVA.Residual_df
 Print "Total SS = "; r.ANOVA.Total_SS
 Print "Total df = "; r.ANOVA.Total_df
 Print "S^2 = "; r.ANOVA.S2
 Print "F = "; r.ANOVA.LINREG_F

 ' enable menu selections
 CIMnu.Enabled = True
 TestSlopeMnu.Enabled = True
 TestInterceptMnu.Enabled = True
 TestR2Mnu.Enabled = True
End Sub

Sub TestInterceptMnu_Click ()
 Dim testValue As Double
 Dim probability As Double
 Dim calcT As Double
```

Listing 10.3 (*Continued*)

```
Dim tableT As Double
Dim passTest As Integer

testValue = 25#
probability = 95
LR_Int_T_test r, probability, testValue, calcT, tableT, _
      passTest
Cls
Print "*********** Test for the Intercept ***********"
Print
Print "At "; probability; "% probability"
Print "H0: "; r.Intercept; "?=? "; testValue;
If passTest Then
   Print "cannot be rejected"
Else
   Print "cannot be accepted"
End If
End Sub


Sub TestR2Mnu_Click ()
 Dim probability As Double
 Dim calcT As Double
 Dim tableT As Double
 Dim passTest As Integer


 probability = 95
 LR_R2_T_Test r, probability, calcT, tableT, passTest
 Cls
 Print "*********** Testing R-squared ************"
 Print
 Print "At "; probability; "% probability"
 Print "H0: "; r.R2; " is 0 ";
 If passTest Then
    Print "cannot be rejected"
 Else
    Print "cannot be accepted"
 End If
End Sub


Sub TestSlopeMnu_Click ()
 Dim testValue As Double
 Dim probability As Double
 Dim calcT As Double
 Dim tableT As Double
 Dim passTest As Integer


 testValue = -1.5
 probability = 95
 LR_Slope_T_test r, probability, testValue, calcT, tableT, _
        passTest
 Cls
 Print "*********** Test for the Slope ***********"
 Print
 Print "At "; probability; "% probability"
 Print "H0: "; r.Slope; "?=? "; testValue;
 If passTest Then
    Print "cannot be rejected"
 Else
    Print "cannot be accepted"
 End If
End Sub
```

The project uses a form that has a simple menu system but no controls. Table 10.1 shows the menu structure and the names of the menu items. The form has the caption "Linear Regression." The menu option Test has six selections to test the various regression-related calculations. Each one of these menu selections clears the form and then displays the result of some regression calculations. Thus, you can zoom in on any method by invoking its related menu selection.

The program tests the various regression-related functions using internal data. The Linear Regression menu selection performs the following tasks:

1. Create the matrix mat and initialize it with values. The matrix has 7 rows and 2 columns.

2. Initialize the LinRegrec-type variable r. This task involves calling the subroutine InitializeLinReg. The arguments for this subroutine-call include the variable r, the indices for the x and y variables (zero and one), the FALSE missing-code flag, and the missing-code value of zero.

3. Accumulate the selected data of matrix mat in the statistical summations of variable r. This task calls subroutine sumLinReg and passes the arguments mat( ), r, and 7 (the number of observations).

4. Perform the basic linear regression calculations by calling the subroutine LinReg. The argument for this subroutine-call is the variable r.

5. Display the regression results. These results include the number of points, the value of R^2, the slope, the intercept, and the ANOVA table components.

6. Enable the remaining menu selections (except the Automatic Best Fit menu selection).

The Confidence Intervals menu selection obtains and displays the 95% confidence interval for the regression coefficients. This task calls subroutine LinRegCoefCI. The arguments for this call are the variables r, probability, slopeHi, slopeLow, intHi, and inLow.

TABLE 10.1 The Menu System for the TSLINREG.MAK Project

Menu caption	Name
&Test	TesMnu
Linear Regression	LRMnu
Confidence Intervals	CIMnu
Test Slope	TestSlopeMnu
Test Intercept	TestInterceptMnu
Test R^2	TestR2Mnu
–	N2
Automatic Best Fit	
–	N1
&Exit	ExitMnu

The Test Slope menu selection tests if the slope is significantly different from –1.5. This task calls the subroutine LR_Slope_T_test and displays the results and inference drawn from the test. The arguments for the subroutine-call are variables r, probability, testValue, calcT, tableT, and passTest.

The Test Intercept menu selection tests if the intercept is significantly different from 25. This task calls the subroutine LR_Int_T_test and displays the results and inference drawn from the test. The arguments for the subroutine-call are variables r, probability, testValue, calcT, tableT, and passTest.

The Test R^2 menu selection tests if the value of R^2 is significantly different from 0. This task calls the subroutine LR_R2_T_test and displays the results and inference drawn from the test. The arguments for the subroutine-call are variables r, probability, calcT, tableT, and passTest.

The Automatic Best Fit menu selection performs the following tasks:

1. Store a new set of data in the local matrix mat. The matrix has five rows and two columns.

2. Perform the automatic best fit by calling the function BestFit. The arguments for this function call are mat( ), 5, numFx, 0, 1, and bestR2( ). The latter argument is the array of the BestLRrec data type.

3. Display the results of the automatic curve fit.

Figure 10.3 shows the output of the test program for each menu selection.

```
************* Linear Regression **********

Number of points: 7
R^2 = 0.98766
Slope = -1.58928571428571
Intercept = 26.9285714285714

************* ANOVA **********
Regression SS = 282.892857142857
Regression df = 1
Residual SS = 3.53571428571433
Residual df = 5
Total SS = 286.428571428571
Total df = 6
S^2 = .707142857142867
F = 400.0505050505


************* Confidence Interval *************

At 95 % probability
Range for slope is -1.79310480790733 to -1.3854666206641
Range for intercept is 25.1055580347702 to 28.7515848223727


********** Test for the Slope ***********

At 95 % probability
```

Figure 10.3 The output of the sample Visual Basic program for linear regression.

```
H0: -1.58928571428571 ?=? -1.5 cannot be rejected


*********** Test for the Intercept ***********

At 95 % probability
H0: 26.9285714285714 ?=? 25 cannot be accepted


*********** Testing R-squared ************

At 95 % probability
H0: .987655860349126 is 0 cannot be accepted


************* Automatic Best Fit *****************

Function #  R2        Slope       Intercept

   2     1.        1.E+00      -4.5552E-06
   3     0.98334   2.9931E-01   1.8194E-01
   0     0.87297   2.2095E-02   2.5328E+00
   1     0.77576   6.2868E-03   9.5309E-01
```

Figure 10.3 (*Continued*)

Multiple and Polynomial Regression

This chapter looks at multiple and polynomial regression as extensions to the linear regression presented in chapter 10. The Visual Basic source code included in this chapter implements the following aspects of multiple and polynomial regression:

- The coefficient of determination

- The regression intercept and slopes

- The standard error for the regression slopes

- The critical Student-t values for the regression slopes

- The regression ANOVA table

Multiple Regression

Multiple regression correlates multiple independent variables with a single dependent variable, as shown in the following equation:

$$Y = A_0 + A_1X_1 + A_2X_2 + ... + A_nX_n \qquad (11.1)$$

where X_1 through X_n are the independent variables (or mathematical transformations of such variables), Y is the dependent variable, A_0 is the intercept, and A_1 through A_n are the regression slopes for the variables X_1 through X_n, respectively. Figure 11.1 shows a sample polynomial regression.

To obtain the intercept and slope for equation 11.1, a program needs to initialize statistical summations for the observed data. This process generates a set of simul-

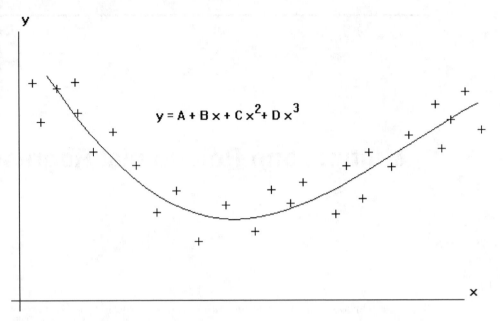

Figure 11.1 A sample polynomial regression.

taneous linear equations that the program must solve to obtain the regression coefficients. Here is the algorithm for obtaining the statistical summations:

Given:

- The matrix of observed data, Mat
- The array of indices, Idx, that selects the columns of the matrix Mat that participate in the regression (The element Idx[0] selects the column that represents the variable Y, the element Idx[1] selects the column that represents the variable X_1, the element Idx[2] selects the column that represents the variable X_2, and so on.)
- The number of observations, NData
- The number of independent variable, N
- The matrix of summations, A, with rows and columns that have indices 0 to N
- The solution vector, B, with indices ranging from 0 to N

Algorithm:

1. Create an array T with indices ranging from 0 to N
2. Assign zeros to the elements of matrix A and vector B and to sumYY (sum of Y squared)
3. For i = 0 to NData – 1, repeat the following steps:
    3.1. For j = 0 to N – 1, set T[j] = Mat(i, Idx[j])
    3.2. Set Yt = T[0]

3.3. Set T[0] = 1

3.4. Add Yt to sumYY

3.5. For k = 0 to N, repeat the following steps:

    3.5.1. Add T[k] * Yt to B[k]

    3.5.2. For j = 0 to N, add T[j] * T[k] to A[j,k]

4. Remove dynamic array T

Here is the algorithm that indicates how to obtain the regression results from the statistical summation matrix A and the solution vector B:

Given:

- The summation matrix, A

- The solution vector, B

- The number of independent variables, N

Algorithm:

1. Set sum = A[0, 0]

2. Set sumY = B[0]

3. For i = 1 to N, repeat the following steps:

    3.1. Set $A[i, 0] = \dfrac{B[i] - sumY * A[0, i]}{sum}$

    3.2. For j = 1 to i, repeat the next steps:

        3.2.1. Subtract $\dfrac{A[0, i] * A[0, j]}{sum}$ from A[j, i]

        3.2.2. Copy A[j, i] into A[j, i]

4. Subtracts $\dfrac{sumY * sumY}{sum}$ from sumYY

5. Set $sumY = \dfrac{sumY}{sum}$

6. For i = 1 to N, repeat the following steps:

    6.1. Set $A[0, i] = \dfrac{A[0, i]}{sum}$

    6.2. Set Mean[i] = A[0, i]

    6.3. Set StdError[i] = A[i, i]

7. Form the correlation matrix by using a For loop with i = 1 to N that repeats the following steps:

    7.1. Set $A[i, 0] = \dfrac{A[i, 0]}{\sqrt{(sumYY * StdError[i])}}$

    7.2. For j = 1 to i − 1 set $A[j, i] = \dfrac{A[j, i]}{\sqrt{(StdError[i] * StdError[j])}}$

    7.3. Set A[i, i] = 1

8. Complete the symmetric part of the matrix by using a For loop with i = 2 to N that repeats the following steps:

    8.1. For j = 1 to i − 1, set A[i, j] = A[j, i]

9. Start solving the simultaneous equations by using a For loop with j = 1 to N:

    9.1. Set Diag = A[j, j]

    9.2. Set A[j, j] = 1

    9.3. For k = 1 to N, set $A[j, k] = \dfrac{A[j, k]}{Diag}$

    9.4. For k = 1 to N, repeat the following statements if j does not equal to k:

        9.4.1. Set temp = A[k, j]

        9.4.2. Set A[k, j] = 0

        9.4.3. For i = 1 to N, subtract A[j, i] * temp from A[k, i]

10. Set R2 = 0

11. Set Intercept = 0

12. For i = 1 to N, repeat the following steps:

    12.1. Set B[i] = 0

    12.2. For j = 1 to N, add A[j, 0] * A[j, i] to B[i]

    12.3. Add B[i] * A[i, 0] to R2

    12.4. Set $B[i] = B[i] * \sqrt{\left(\dfrac{sumYY}{StdError[i]}\right)}$

    12.5. Add B[i] * A[0, i] to Intercept

13. Set Intercept = sumY − Intercept

14. Set DegF = sum − N − 1

15. Set $MS = (1 - R2) * \dfrac{sumYY}{DegF}$

16. For i = 1 to N, repeat the following steps:

    16.1. Set $StdErSlope[i] = \sqrt{\left(\dfrac{MS * A(i, i)}{StdError[i]}\right)}$

    16.2. Set $TCalc[i] = \dfrac{B[i]}{StdError[i]}$

17. Set $R2adj = 1 - (1 - R2) * \dfrac{sum}{DegF}$

18. Set Total_df = N + 1

19. Set Total_SS = sumYY

20. Set Reg_df = N

21. Set S2 = MS;

22. Set Reg_SS = sumYY − Set S2

23. Set Residual_df = DegF

24. Set Residual_SS = S2 * DegF

25. Set $F = \dfrac{DegF}{N} * \dfrac{R2}{(1 - R2)}$

The items calculated in steps 18 to 25 form the entries in the regression ANOVA table. The item R2adj represents the adjusted value for the coefficient of correlation. The adjustment takes into account the number of terms and the number of observations. You can then compare the adjusted value with similar ones obtained from other multiple regressions with different numbers of terms and observations.

The Visual Basic source code

Listing 11.1 shows the source code for the MLREG.BAS module file. This file declares two data types, MLR_ANOVA and MLRegrec. The data type MLR_ANOVA, as the name might suggest, contains fields that support the ANOVA table for the multiple regression. The data type MLRegrec contains fields that support the multiple regression calculations. The data type contains the following groups of fields:

- The field numTerms stores the number of independent variables involved in the multiple regression.

- The fields sum, sumY, and sumYY store the number of observations, sum of Y, and sum of Y squared, respectively.

- The fields hasMissingData and missingCode are the missing-code flag and missing-code value, respectively.

- The field Inverted_S is a flag that determines whether or not the field S contains an inverted matrix.

- The field Intercept stores the intercept of the multiple regression.

- The fields R2 and R2adj store the coefficient of correlation and its adjusted value, respectively.

- The field ANOVA stores the ANOVA table for the multiple regression.

Throughout the book, the underscore character is used to split wrapping lines of Visual Basic declarations and statements.

Listing 11.1 The source code for the MLREG.BAS module file.

```
Global Const MLREG_BIG# = 1E+30

Type MLR_ANOVA
 Reg_df As Double
 Reg_SS As Double
 Residual_df As Double
 Residual_SS As Double
 Total_df As Double
 Total_SS As Double
 S2 As Double
 F As Double
End Type
```

Listing 11.1 (*Continued*)

```
Type MLRegrec
 numTerms As Integer
 ' summation block
 sum As Double
 sumY As Double
 sumYY As Double
 hasMissingData As Integer
 missingCode As Double
 Inverted_S As Integer
 Intercept As Double
 R2 As Double
 R2adj As Double
 ANOVA As MLR_ANOVA
End Type

Sub CalcMultiReg (r As MLRegrec, A() As Double, B() As Double, _
 S() As Double, Mean() As Double, StdErSlope() As Double, _
 TCalc() As Double)

 Dim Diag As Double, tempo As Double
 Dim DegF As Double, MS As Double
 Dim i As Integer, j As Integer
 Dim n As Integer, k As Integer

 n = r.numTerms + 1
 r.sum = A(0, 0)
 r.sumY = B(0)
 ' Form the A and S Matrices
 For i = 1 To n - 1
  A(i, 0) = B(i) - r.sumY * A(0, i) / r.sum
  For j = 1 To i
   A(j, i) = A(j, i) - A(0, i) * A(0, j) / r.sum
   ' Make a copy of the matrix A
   S(j, i) = A(j, i)
  Next j
 Next i

 ' Clear the matrix S inversion flag
 r.Inverted_S = False

 r.sumYY = r.sumYY - r.sumY ^ 2 / r.sum
 r.sumY = r.sumY / r.sum
 For i = 1 To n - 1
  A(0, i) = A(0, i) / r.sum
  ' Make copies of the mean and std error vectors
  Mean(i) = A(0, i)
  StdErSlope(i) = A(i, i)
 Next i

 ' Form correlation matrix
 For i = 1 To n - 1
  A(i, 0) = A(i, 0) / Sqr(r.sumYY * StdErSlope(i))
  For j = 1 To i - 1
   A(j, i) = A(j, i) / Sqr(StdErSlope(i) * StdErSlope(j))
  Next j
  A(i, i) = 1#
 Next i

 ' Complete symmetric part of the matrix
 For i = 2 To n - 1
  For j = 1 To i - 1
   A(i, j) = A(j, i)
  Next j
 Next i
```

Listing 11.1 (*Continued*)

```
' Start solving the simultaneous equations
For j = 1 To n - 1
 Diag = A(j, j)
 A(j, j) = 1
 For k = 1 To n - 1
  A(j, k) = A(j, k) / Diag
 Next k

 For k = 1 To n - 1
  If j <> k Then
   tempo = A(k, j)
   A(k, j) = 0
   For i = 1 To n - 1
    A(k, i) = A(k, i) - A(j, i) * tempo
   Next i
  End If
 Next k
Next j
r.R2 = 0
r.Intercept = 0
For i = 1 To n - 1
 B(i) = 0
 For j = 1 To n - 1
  B(i) = B(i) + A(j, 0) * A(j, i)
 Next j
 r.R2 = r.R2 + B(i) * A(i, 0)
 B(i) = B(i) * Sqr(r.sumYY / StdErSlope(i))
 r.Intercept = r.Intercept + B(i) * A(0, i)
Next i
r.Intercept = r.sumY - r.Intercept
DegF = r.sum - r.numTerms - 1
MS = (1 - r.R2) * r.sumYY / DegF
For i = 1 To n - 1
 StdErSlope(i) = Sqr(MS * A(i, i) / StdErSlope(i))
 TCalc(i) = B(i) / StdErSlope(i)
Next i
r.R2adj = 1 - (1 - r.R2) * r.sum / DegF
r.ANOVA.Total_df = r.numTerms + 1
r.ANOVA.Total_SS = r.sumYY
r.ANOVA.Reg_df = r.numTerms
r.ANOVA.S2 = MS
r.ANOVA.Reg_SS = r.sumYY - r.ANOVA.S2
r.ANOVA.Residual_df = DegF
r.ANOVA.Residual_SS = r.ANOVA.S2 * DegF
r.ANOVA.F = DegF / r.numTerms * r.R2 / (1 - r.R2)
End Sub

Sub CalcSumMLR (DataMat() As Double, r As MLRegrec, _
NData As Integer, A() As Double, B() As Double, _
Index() As Integer)
Dim i As Integer, j As Integer
Dim n As Integer, k As Integer
Dim Ttrnsf() As Double
Dim Torig() As Double
Dim Yt As Double
Dim OK As Integer

n = r.numTerms + 1
ReDim Ttrnsf(n)
ReDim Torig(n)

For i = 0 To NData - 1
 ' copy data to local array Torig
 For j = 0 To n - 1
```

Listing 11.1 (*Continued*)

```
    k = Index(j)
    Torig(j) = DataMat(i, k)
  Next j

  ' set ok flag
  OK = True

  ' are there any possible missing data?
  If r.hasMissingData Then
   ' search for missing data
   j = 0
   Do While (j < n) And OK
     If Torig(j) > r.missingCode Then
       j = j + 1
     Else
       OK = False
     End If
   Loop
  End If

  If OK Then
     ' now transform the data
     ' first transform Torig() into Ttrnsf()
     For j = 0 To n - 1
       Ttrnsf(j) = MyMLRFx(Torig(j), j)
     Next j

     Yt = Ttrnsf(0)
     Ttrnsf(0) = 1
     r.sumYY = r.sumYY + Yt ^ 2
     For k = 0 To n - 1
       B(k) = B(k) + Ttrnsf(k) * Yt
       For j = 0 To n - 1
         A(j, k) = A(j, k) + Ttrnsf(j) * Ttrnsf(k)
       Next j
     Next k
  End If
 Next i
End Sub

Sub InitializeMLR (r As MLRegrec, numTerms As Integer, _
 hasMissingData As Integer, missingCode As Double, _
 A() As Double, B() As Double)

 Dim i As Integer
 Dim j As Integer
 Dim n As Integer

 r.numTerms = numTerms
 n = r.numTerms + 1

 For i = 0 To n - 1
   ' initialize summations
   For j = 0 To n - 1
     A(i, j) = 0
   Next j
   B(i) = 0
 Next i
 r.sumYY = 0
 r.hasMissingData = hasMissingData
 r.missingCode = missingCode
End Sub
```

Listing 11.1 (*Continued*)

```
Sub MLR_R2_T_Test (r As MLRegrec, probability As Double, _
 calcT As Double, tableT As Double, passTest As Integer)
' test hypothesis H0 : R^2 = 0

 Dim p As Double, df As Double

 If probability > 1 Then
  p = .5 - probability / 200
 Else
  p = .5 - probability / 2
 End If
 df = r.sum - (r.numTerms + 1)
 tableT = TInv(p, df)
 calcT = Sqr(r.R2 * df / (1 - r.R2))
 passTest = calcT <= tableT
End Sub


Sub MLR_Slope_T_test (r As MLRegrec, probability As Double, _
 testValue As Double, termNum As Integer, calcT As Double, _
 tableT As Double, passTest As Integer, B() As Double, _
 StdErSlope() As Double)
 Dim p As Double, df As Double

 If probability > 1 Then
  p = .5 - probability / 200
 Else
  p = .5 - probability / 2
 End If
 df = r.sum - (r.numTerms + 1)
 tableT = TInv(p, df)
 calcT = (B(termNum) - testValue) / StdErSlope(termNum)
 passTest = Abs(calcT) <= tableT
End Sub


Sub MultiRegCoefCI (r As MLRegrec, probability As Double, _
 slopeHi() As Double, slopeLow() As Double, B() As Double, _
 StdErSlope() As Double)

 Dim diff As Double, p As Double
 Dim df As Double, tableT As Double
 Dim j As Integer
 Dim n As Integer
 n = r.numTerms + 1
 If probability > 1 Then
  p = .5 - probability / 200
 Else
  p = .5 - probability / 2
 End If
 df = r.sum - (r.numTerms + 1)
 tableT = TInv(p, df)
 For j = 1 To n - 1
  diff = tableT * StdErSlope(j)
  slopeHi(j) = B(j) + diff
  slopeLow(j) = B(j) - diff
 Next j
End Sub


Sub yHatCI (r As MLRegrec, probability As Double, _
 X() As Double, yHatLow As Double, yHat As Double, _
 yHatHigh As Double, B() As Double, S() As Double, _
 Mean() As Double)
```

Listing 11.1 (*Continued*)

```
Dim diff As Double, p As Double
Dim df As Double, tableT As Double
Dim traceMat As Double, pivot As Double
Dim tempo As Double
Dim i As Integer, j As Integer
Dim k As Integer, m As Integer
Dim XX() As Double
Dim prodMat() As Double
Dim Xcopy() As Double
Dim n As Integer

n = r.numTerms + 1
ReDim XX(n, n)
ReDim prodMat(n, n)
ReDim Xcopy(n)

If probability > 1 Then
 p = .5 - probability / 200
Else
 p = .5 - probability / 2
End If
df = r.sum - (r.numTerms + 1)
tableT = TInv(p, df)

If Not r.Inverted_S Then ' Invert matrix S
 r.Inverted_S = True
 For j = 1 To n - 1
  pivot = S(j, j)
  S(j, j) = 1
  For k = 1 To n - 1
   S(j, k) = S(j, k) / pivot
  Next k
  For k = 1 To n - 1
   If k <> j Then
    tempo = S(k, j)
    S(k, j) = 0
    For m = 1 To n - 1
     S(k, m) = S(k, m) - S(j, m) * tempo
    Next m
   End If
  Next k
 Next j
End If

' Calculate yHat
yHat = r.Intercept
For i = 1 To n - 1
 Xcopy(i) = MyMLRFx(X(i), i)
 yHat = yHat + B(i) * Xcopy(i)
Next i

' Form standarized vector
For i = 1 To n - 1
 Xcopy(i) = Xcopy(i) - Mean(i)
Next i

' Form X X' matrix
For k = 1 To n - 1
 For j = 1 To n - 1
  XX(j, k) = 0
 Next j
Next k
```

Listing 11.1 (*Continued*)

```
For k = 1 To n - 1
 For j = 1 To n - 1
  XX(j, k) = XX(j, k) + Xcopy(j) * Xcopy(k)
 Next j
Next k

' Multiply S_Inverse and XX'
For i = 1 To n - 1
 For j = 1 To n - 1
  prodMat(i, j) = 0
  For k = 1 To n - 1
   prodMat(i, j) = prodMat(i, j) + S(i, k) * XX(k, j)
  Next k
 Next j
Next i

' Calculate trace of prodMat
traceMat = 1#
For i = 1 To n - 1
 traceMat = traceMat * prodMat(i, i)
Next i

diff = tableT * Sqr(r.ANOVA.S2 * (1 + 1 / r.sum + traceMat))
yHatLow = yHat - diff
yHatHigh = yHat + diff

yHat = MyMLRInvFy(yHat)
yHatLow = MyMLRInvFy(yHatLow)
yHatHigh = MyMLRInvFy(yHatHigh)
End Sub
```

The module file declares the following routines:

1. The subroutine InitializeMLR initializes a MLRegrec-type variable to prepare it for a new set of calculations. The parameter r is the initialized MLRegrec-type variable. The parameter NumTerms specifies the number of independent variables. The parameter HasMissingData specifies the missing-code flag. The parameter MissingCode indicates the numeric value of the missing code. If the argument to HasMissingData is zero, the argument for MissingCode can be any value. The parameters A( ) and B( ) pass the matrix and solution vectors, respectively, to be initialized by the subroutine. You must call subroutine InitializeMLR at least once before you perform multiple regression analysis.

2. The subroutine CalcSumMLR updates the statistical summations. The parameter DataMat( ) is a matrix that supplies the observed data. The parameter r is the MLRegrec-type variable that stores the statistical summations. The parameter NData indicates the number of observations to process. The parameters A( ), B( ), and Index( ) pass the summations matrix, summations vector, and the variable-selection index array, respectively. You can call subroutine CalcSumMLR more than once before calling the next function.

3. The subroutine CalcMultiReg calculates the regression slopes, intercept, ANOVA, and other related data. The parameter r is the MLRegrec-type variable that holds the regression matrices and vectors. The parameters A( ), B( ), S( ), Mean( ),

StdErSlope( ), and TCalc( ) pass the summations matrix, summations vector, inverse summations natrix, means array, standard slope errors array, and calculated Student-t values array, respectively. You must call subroutine CalcMultiReg before calling the remaining functions listed here.

4. The subroutine yHatCI calculates the projected value of the dependent variable and its confidence interval for a given combination of independent variables. The parameter r is the MLRegrec-type variable that contains the results of multiple regression. The parameter Probability specifies the confidence level (expressed as a percentage or a decimal). The parameter X( ) specifies the array that stores the independent variables: the element X(1) stores the value for X1, the element X(2) stores the value for X2, and so on. The parameters yHat, yHatLow, and yHatHigh report back the value for the projected value and its confidence interval. The parameters B( ), S( ), and Mean( ) pass the summations vector, inverse summations matrix, and means array, respectively.

5. The subroutine MultiRegCoefCI calculates the confidence intervals for the regression slopes. The parameter r is the MLRegrec-type variable that contains the results of multiple regression. The parameter Probability specifies the confidence level (expressed as a percentage or a decimal). The parameter slopeHi( ) specifies the array of upper limits for the regression slope: the element slopeHi(1) corresponds to the slope for variable X1, the element slopeHi(2) corresponds to the slope for variable X2, and so on. The parameter slopeLow( ) specifies the array of lower limits for the regression slope: the element slopeLow(1) corresponds to the slope for variable X1, the element slopeLow(2) corresponds to the slope for variable X2, and so on. The parameters B( ) and StdErSlope( ) pass the summations vector and the array of standard slope errors, respectively.

6. The subroutine MLR_Slope_T_test tests the value of a specific regression slope. The parameter r is the MLRegrec-type variable that contains the results of multiple regression. The parameter Probability specifies the confidence level (expresses either as a percent or a decimal). The parameter testValue specifies the tested value of the targeted regression slope. The parameter termNum is the index of the tested regression slope. The parameter calcT passes back the calculated Student-t value. The parameter tableT passes back the tabulated Student-t value. The parameter passTest is a Boolean flag that reports whether the outcome of the test is positive or negative.

7. The subroutine MLR_R2_T_test tests whether the value of the correlation coefficient is not zero. The parameter r is the MLRegrec structure that contains the results of multiple regression. The parameter Probability specifies the confidence level (expressed as a percent or a decimal). The parameter calcT reports back the calculated Student-t value. The parameter tableT reports back the tabulated Student-t value. The parameter passTest is a flag that reports back whether the outcome of the test is positive or negative. The parameters B( ) and StdErSlope( ) pass the summations vector and the array of standard slope errors, respectively.

Listing 11.2 shows the source code for the MYMLREG.BAS module file. The file contains the Visual Basic functions MyMLRFx and MyMLRInvFy. The first Visual Basic function transforms the data (both the dependent variable and independent variables). The function uses the parameter X and Index. The second parameter acts as a selector. The function uses the Select Case statement to examine the value of parameter Index and accordingly perform the required transformations. The Case 0 label deals with transforming the values of Y. Typically, the function uses Case 1, Case 2, and Case 3 labels (and so on) to transform variables X1, X2, and X3 (and so on). The current form merely uses the Case Else label to return the value of parameter X unchanged. Thus, the current form of function MyMLRFx performs no nonlinear transformation of the data.

Listing 11.2   The source code for the MYMLREG.BAS module file.

```
Function MyMLRFx (X As Double, Index As Integer)
 Select Case Index
  Case 0 ' Y
   MyMLRFx = X

  Case Else
   MyMLRFx = X
 End Select
End Function

Function MyMLRInvFy (Y As Double)
 MyMLRInvFy = Y
End Function
```

The Visual Basic test program

Let's look at the test program for the multiple regression library. Listing 11.3 shows the source code for the form associated with the TSMLREG.MAK program project. To compile the test program, you need to include the files MLRREG.BAS, MYMLREG.BAS, and STATLIB.BAS in your project file.

Listing 11.3   The source code for the form associated with the TSMLREG.MAK program project.

```
Dim A(10, 10) As Double
Dim S(10, 10) As Double
Dim B(10) As Double
Dim Mean(10) As Double
Dim StdErSlope(10) As Double
Dim TCalc(10) As Double
Dim Index(10) As Integer
Dim mat(15, 10) As Double

Dim r As MLRegrec

Dim numTerms As Integer

Sub ExitMnu_Click ()
 End
End Sub

Sub Form_Load ()
 SlopeCIMnu.Enabled = False
 TestSlopesMnu.Enabled = False
 TestR2Mnu.Enabled = False
 YhatMnu.Enabled = False
End Sub
```

Listing 11.3 *(Continued)*

```
Sub MLRMnu_Click ()
 Dim i As Integer
 Dim numData As Integer

 mat(0, 0) = 162
 mat(1, 0) = 120
 mat(2, 0) = 223
 mat(3, 0) = 131
 mat(4, 0) = 67
 mat(5, 0) = 169
 mat(6, 0) = 81
 mat(7, 0) = 192
 mat(8, 0) = 116
 mat(9, 0) = 55
 mat(10, 0) = 252
 mat(11, 0) = 232
 mat(12, 0) = 144
 mat(13, 0) = 103
 mat(14, 0) = 212

 mat(0, 1) = 274
 mat(1, 1) = 180
 mat(2, 1) = 375
 mat(3, 1) = 205
 mat(4, 1) = 86
 mat(5, 1) = 265
 mat(6, 1) = 98
 mat(7, 1) = 330
 mat(8, 1) = 195
 mat(9, 1) = 53
 mat(10, 1) = 430
 mat(11, 1) = 372
 mat(12, 1) = 236
 mat(13, 1) = 157
 mat(14, 1) = 370

 mat(0, 2) = 2450
 mat(1, 2) = 3254
 mat(2, 2) = 3802
 mat(3, 2) = 2838
 mat(4, 2) = 2347
 mat(5, 2) = 3782
 mat(6, 2) = 3008
 mat(7, 2) = 2450
 mat(8, 2) = 2137
 mat(9, 2) = 2560
 mat(10, 2) = 4020
 mat(11, 2) = 4427
 mat(12, 2) = 2660
 mat(13, 2) = 2088
 mat(14, 2) = 2605

 numData = 15
 numTerms = 2

 Index(0) = 0
 Index(1) = 1
 Index(2) = 2
 Index(3) = 3

 InitializeMLR r, numTerms, False, 0, A(), B()
 CalcSumMLR mat(), r, numData, A(), B(), Index()
 CalcMultiReg r, A(), B(), S(), Mean(), StdErSlope(), TCalc()
```

Listing 11.3 *(Continued)*

```
Cls
Print "************* Multiple Regression **********"
Print
Print "Number of points: "; Format$(r.sum, "###")
Print "R^2 = "; Format$(r.R2, "0.00000")
Print "Adjusted R^2 = "; Format$(r.R2adj, "0.00000")
Print "Intercept = "; Format$(r.Intercept, "#.####E+00")
For i = 1 To numTerms
 Print "Slope for X";
 Print Format$(i, "#");
 Print " = ";
 Print Format$(B(i), "#.####E+00")
Next i
Print
Print "************* ANOVA **********"
Print "Regression SS = "; Format$(r.ANOVA.Reg_SS, "#.####E+00")
Print "Regression df = "; Format$(r.ANOVA.Reg_df, "#.####E+00")
Print "Residual SS = "; _
 Format$(r.ANOVA.Residual_SS, "#.####E+00")
Print "Residual df = "; _
 Format$(r.ANOVA.Residual_df, "#.####E+00")
Print "Total SS = "; Format$(r.ANOVA.Total_SS, "#.####E+00")
Print "Total df = "; Format$(r.ANOVA.Total_df, "#.####E+00")
Print "S^2 = "; Format$(r.ANOVA.S2, "#.####E+00")
Print "F = "; Format$(r.ANOVA.F, "#.####E+00")

 ' enable remaining menu selections
 SlopeCIMnu.Enabled = True
 TestSlopesMnu.Enabled = True
 TestR2Mnu.Enabled = True
 YhatMnu.Enabled = True
End Sub

Sub SlopeCIMnu_Click ()
 Static slopeLow(10) As Double
 Static slopeHi(10) As Double
 Dim probability As Double

 probability = 95
 MultiRegCoefCI r, probability, slopeHi(), slopeLow(), _
         B(), StdErSlope()
 Cls
 Print "************* Slope Confidence Intervals *************"
 Print
 Print "At "; Format$(probability, "##.0"); "% probability"
 For i = 1 To numTerms
  Print "Range for slope of X";
  Print Format$(i, "#");
  Print " is "; Format$(slopeLow(i), "0.00000E+00");
  Print " to "; Format$(slopeHi(i), "0.00000E+00")
 Next i
End Sub

Sub TestR2Mnu_Click ()
 Dim probability As Double
 Dim testValue As Double
 Dim passTest As Integer
 Dim calcT As Double
 Dim tableT As Double
 probability = 95
 Cls
 Print "************ Testing R^2 *************"
 Print
```

Listing 11.3 *(Continued)*

```
  Print "At "; Format$(probability, "##.0"); "% probability"
  MLR_R2_T_Test r, probability, calcT, tableT, passTest
  Print "H0: ";
  Print Format$(r.R2, "0.000000");
  Print " is not 0, ";
  If passTest Then
    Print "cannot be rejected"
  Else
    Print "cannot be accepted"
  End If
End Sub

Sub TestSlopesMnu_Click ()
 Dim i As Integer
 Dim probability As Double
 Dim testValue As Double
 Dim passTest As Integer
 Dim calcT As Double
 Dim tableT As Double

 probability = 95
 Cls
 Print "************ Testing Slopes *************"
 Print
 Print "At "; Format$(probability, "##.0"); "% probability"
 For i = 1 To numTerms
  Select Case i
   Case 1
    testValue = .5

   Case 2
    testValue = .01
  End Select

  MLR_Slope_T_test r, probability, testValue, i, calcT, _
         tableT, passTest, B(), StdErSlope()
  Print "H0: "; Format$(B(i), "0.00000E+00");
  Print " ?=? "; testValue;
  If passTest Then
    Print " cannot be rejected"
  Else
    Print " cannot be accepted"
  End If
 Next i
End Sub

Sub YhatMnu_Click ()
 Static Xarr(10) As Double
 Dim i As Integer
 Dim probability As Double
 Dim Yhat As Double
 Dim YHatLow As Double
 Dim YHatHigh As Double

 For i = 1 To numTerms
  Xarr(i) = mat(0, i)
 Next i

 probability = 95
 yHatCI r, probability, Xarr(), YHatLow, Yhat, YHatHigh, _
     B(), S(), Mean()
 Cls
 Print "************ Projected Y *************"
 Print
```

Listing 11.3 (*Continued*)

```
Print "At "; Format$(probability, "##.0"); "% probability"
For i = 1 To numTerms
 Print "X(";
 Print Format$(i, "#");
 Print ") = "; Format$(Xarr(i), "#####.##")
Next i
Print "Y^ = "; Format$(Yhat, "0.00000E+00")
Print "Range for Y^ is ";
Print Format$(YHatLow, "0.00000E+00");
Print " to "";
Print Format$(YHatHigh, "0.00000E+00")
End Sub
```

The project uses a form that has a simple menu system but no controls. Table 11.1 shows the menu structure and the names of the menu items for project TSML-REG.MAK. The form has the caption "Multiple Regression." The menu option Test has five selections to test the various regression-related calculations. Each one of these menu selections clears the form and then displays the result of some regression calculations. Thus, you can zoom in on any method by invoking its related menu selection.

The program tests the various regression-related functions using internal data. The test program declares a collection of arrays, matrices, a MLRegrec-type variable, and a simple variable in the form level. The Multiple Regression menu selection performs the following tasks:

1. Assign the test data to the elements of matrix mat.

2. Assign the values to the variables that store the number of data and the number of terms (that is, the number of independent variables).

3. Assign the indices for the regression variables to the elements of array Index.

4. Initialize the MLRegrec-type variable r by calling the subroutine InitializeMLR. The arguments for the subrotuine call are the variable r, the variable numTerms, the array Index, the value FALSE, the value 0, the matrix A, and the array B. The latter two arguments indicate that the data matrix will have no missing observations.

5. Update the statistical summations by calling subroutine CalcSumMLR. The arguments for this function call are the matrix Mat, the variable r, the variable numData, the matrix A, the array B, and the array Index.

6. Calculate the regression slopes and other results by calling the subroutine Calc-MultiReg. The argument for this subroutine call is the variable r, the matrix A, the array B, the array Mean, the array StdErSlope, and the array TCalc.

7. Display the regression results, including the coefficient of correlation, the intercept, the regression slopes, and the ANOVA table components.

8. Enable the remaining menu selections.

The Slope Confidence Intervals menu selection calculates and displays the confidence intervals, at 95% confidence, for the regression slopes. This task involves calling subroutine MultiRegCoefCI. The program uses the arrays slopeHi and slopeLow to obtain the sought confidence intervals.

TABLE 11.1 The Menu System for the TSMLREG.MAK Project

Menu caption	Name
&Test	TesMnu
Multiple Regression	MLRMnu
Slope Confidence Intervals	SlopeCIMnu
Test Slopes	TestSlopesMnu
Test R^2	TestR2Mnu
Projected Y	YhatMnu
–	N1
&Exit	ExitMnu

The Test Slopes menu selection tests the values for the two regression slopes at 95% confidence. The program calls subroutine MLR_Slope_T_test to determine if the first and second slopes are not that significantly different from the values of 0.5 and 0.01, respectively. The program displays the outcome of the test.

The Test R^2 menu selection tests whether or not the value of the correlation coefficient is zero (at 95% confidence). The program calls subroutine MLR_R2_T_test and displays the outcome of the test.

The Projected Y menu selection calculates the projected value of the dependent variable and its confidence interval (at 95% confidence) for the first row of data. The program calls subroutine yHatCI and passes the arguments r, probability, Xarr( ), yHatLow, yHat, yHatHigh, B( ), S( ), and Mean( ). The program then displays the value of the dependent variables, the projected value, and the confidence interval for the projection.

Figure 11.2 shows the output of the test program for each menu selection.

Polynomial Regression

Polynomial regression is a special case of multiple regression. Instead of dealing with multiple variables, you correlate two variables using a polynomial model. Thus polynomial regression takes the independent variable X and creates additional pseudo-variables whose values are based on the integer powers of X. Therefore creating a library for polynomial regression is very easy when you start with a library that supports multiple regression. Thus, the multiple regression variables X1, X2, X3, and so on, correspond to X, X^2, X^3, and so on, in polynomial regression.

The Visual Basic source code

Listing 11.4 shows the source code for the POLYREG.BAS module file. This file declares two data types, PolyReg_ANOVA and PolyRegrec. The data type PolyReg_ANOVA, as the name might suggest, contains fields that support the ANOVA table for the polynomial regression. The data type PolyRegrec contains fields that support the polynomial regression calculations. The structure contains the following groups of fields:

- The field numTerms stores the number of independent variables involved in the polynomial regression.

- The fields sum, sumY, and sumYY store the number of observations, sum of Y, and sum of Y squared, respectively.

- The fields hasMissingData and missingCode are the missing-code flag and missing-code value, respectively.

- The field Inverted_S is a flag that determines whether or not the field S contains an inverted matrix.

- The field Intercept stores the intercept of the polynomial regression.

- The fields R2 and R2adj store the coefficient of correlation and its adjusted value, respectively.

- The field ANOVA stores the ANOVA table for the polynomial regression.

```
************ Multiple Regression **********

Number of points: 15
R^2 = 0.99894
Adjusted R^2 = 0.99868
Intercept = 3.4526E+00
Slope for X1 = 4.96E-01
Slope for X2 = 9.1991E-03

************ ANOVA **********
Regression SS = 5.3897E+04
Regression df = 2.E+00
Residual SS = 5.6884E+01
Residual df = 1.2E+01
Total SS = 5.3902E+04
Total df = 3.E+00
S^2 = 4.7403E+00
F = 5.6795E+03


************ Slope Confidence Intervals ************

At 95.0% probability
Range for slope of X1 is 4.82835E-01 to 5.09175E-01
Range for slope of X2 is 7.09318E-03 to 1.13050E-02


*********** Testing Slopes ************

At 95.0% probability
H0: 4.96005E-01 ?=? .5 cannot be rejected
H0: 9.19908E-03 ?=? .01 cannot be rejected


*********** Testing R^2 ************

At 95.0% probability
H0: 0.998945 is not 0, cannot be accepted


*********** Projected Y ************

At 95.0% probability
X(1) = 274.
X(2) = 2450.
Y^ = 1.61896E+02
Range for Y^ is 1.57003E+02 to 1.66788E+02
```

Figure 11.2 The output of the sample Visual Basic program for multiple regression.

Listing 11.4 The source code for the POLYREG.BAS module file.

```
Global Const POLYREG_BIG# = 1E+30

Type PolyReg_ANOVA
 Reg_df As Double
 Reg_SS As Double
 Residual_df As Double
 Residual_SS As Double
 Total_df As Double
 Total_SS As Double
 S2 As Double
 F As Double
End Type

Type PolyRegrec
 NumTerms As Integer
 ' summation block
 sum As Double
 sumY As Double
 sumYY As Double
 hasMissingData As Integer
 missingCode As Double
 Inverted_S As Integer
 Intercept As Double
 R2 As Double
 R2adj As Double
 ANOVA As PolyReg_ANOVA
End Type

Sub CalcPolyReg (r As PolyRegrec, A() As Double, B() As Double, _
 S() As Double, Mean() As Double, StdErSlope() As Double, _
 TCalc() As Double)

 Dim Diag As Double, tempo As Double
 Dim DegF As Double, MS As Double
 Dim i As Integer, j As Integer
 Dim n As Integer, k As Integer

 n = r.NumTerms + 1
 r.sum = A(0, 0)
 r.sumY = B(0)
 ' Form the A and S Matrices
 For i = 1 To n - 1
  A(i, 0) = B(i) - r.sumY * A(0, i) / r.sum
  For j = 1 To i
   A(j, i) = A(j, i) - A(0, i) * A(0, j) / r.sum
   ' Make a copy of the matrix A
   S(j, i) = A(j, i)
  Next j
 Next i

 ' Clear the matrix S inversion flag
 r.Inverted_S = False

 r.sumYY = r.sumYY - r.sumY ^ 2 / r.sum
 r.sumY = r.sumY / r.sum
 For i = 1 To n - 1
  A(0, i) = A(0, i) / r.sum
  ' Make copies of the mean and std error vectors
  Mean(i) = A(0, i)
  StdErSlope(i) = A(i, i)
 Next i

 ' Form correlation matrix
 For i = 1 To n - 1
```

Listing 11.4 *(Continued)*

```
  A(i, 0) = A(i, 0) / Sqr(r.sumYY * StdErSlope(i))
  For j = 1 To i - 1
   A(j, i) = A(j, i) / Sqr(StdErSlope(i) * StdErSlope(j))
  Next j
  A(i, i) = 1#
Next i

' Complete symmetric part of the matrix
For i = 2 To n - 1
 For j = 1 To i - 1
  A(i, j) = A(j, i)
 Next j
Next i

' Start solving the simultaneous equations
For j = 1 To n - 1
 Diag = A(j, j)
 A(j, j) = 1
 For k = 1 To n - 1
  A(j, k) = A(j, k) / Diag
 Next k

 For k = 1 To n - 1
  If j <> k Then
   tempo = A(k, j)
   A(k, j) = 0
   For i = 1 To n - 1
    A(k, i) = A(k, i) - A(j, i) * tempo
   Next i
  End If
 Next k
Next j
r.R2 = 0
r.Intercept = 0
For i = 1 To n - 1
 B(i) = 0
 For j = 1 To n - 1
  B(i) = B(i) + A(j, 0) * A(j, i)
 Next j
 r.R2 = r.R2 + B(i) * A(i, 0)
 B(i) = B(i) * Sqr(r.sumYY / StdErSlope(i))
 r.Intercept = r.Intercept + B(i) * A(0, i)
Next i
r.Intercept = r.sumY - r.Intercept
DegF = r.sum - r.NumTerms - 1
MS = (1 - r.R2) * r.sumYY / DegF
For i = 1 To n - 1
 StdErSlope(i) = Sqr(MS * A(i, i) / StdErSlope(i))
 TCalc(i) = B(i) / StdErSlope(i)
Next i
r.R2adj = 1 - (1 - r.R2) * r.sum / DegF
r.ANOVA.Total_df = r.NumTerms + 1
r.ANOVA.Total_SS = r.sumYY
r.ANOVA.Reg_df = r.NumTerms
r.ANOVA.S2 = MS
r.ANOVA.Reg_SS = r.sumYY - r.ANOVA.S2
r.ANOVA.Residual_df = DegF
r.ANOVA.Residual_SS = r.ANOVA.S2 * DegF
r.ANOVA.F = DegF / r.NumTerms * r.R2 / (1 - r.R2)
End Sub


Sub CalcSumPolyReg (DataMat() As Double, r As PolyRegrec, _
 NData As Integer, A() As Double, B() As Double, _
 Index() As Integer)
```

Listing 11.4 (*Continued*)

```
Dim i As Integer, j As Integer
Dim n As Integer, k As Integer
Static T(2) As Double
Dim Yt As Double
Dim Xt As Double
Dim OK As Integer

n = r.NumTerms + 1

For i = 0 To NData - 1
 T(0) = DataMat(i, Index(0))
 T(1) = DataMat(i, Index(1))

 ' set ok flag
 OK = True

 ' are there any possible missing data?
 If r.hasMissingData Then
  ' search for missing data
  j = 0
  Do While (j < 2) And OK
   If T(j) > r.missingCode Then
    j = j + 1
   Else
    OK = False
   End If
  Loop
 End If

 If OK Then
    ' now transform the data
    T(0) = MyPolyFx(T(0))
    T(1) = MyPolyFy(T(1))

    Yt = T(0)
    Xt = T(1)
    r.sumYY = r.sumYY + Yt ^ 2
    For k = 0 To n - 1
     B(k) = B(k) + Yt * Xt ^ k
     For j = 0 To n - 1
      A(j, k) = A(j, k) + Xt ^ (j + k)
     Next j
    Next k
 End If
Next i
End Sub

Sub InitializePolyReg (r As PolyRegrec, NumTerms As Integer, _
 hasMissingData As Integer, missingCode As Double, _
 A() As Double, B() As Double)

 Dim i As Integer
 Dim j As Integer
 Dim n As Integer

 r.NumTerms = NumTerms
 n = r.NumTerms + 1

 For i = 0 To n - 1
  ' initialize summations
  For j = 0 To n - 1
   A(i, j) = 0
  Next j
```

Listing 11.4 *(Continued)*

```
  B(i) = 0
 Next i
 r.sumYY = 0
 r.hasMissingData = hasMissingData
 r.missingCode = missingCode
End Sub

Sub PolyReg_R2_T_Test (r As PolyRegrec, probability As Double, _
 calcT As Double, tableT As Double, passTest As Integer)
' test hypothesis H0 : R^2 = 0

 Dim p As Double, df As Double

 If probability > 1 Then
  p = .5 - probability / 200
 Else
  p = .5 - probability / 2
 End If
 df = r.sum - (r.NumTerms + 1)
 tableT = TInv(p, df)
 calcT = Sqr(r.R2 * df / (1 - r.R2))
 passTest = calcT <= tableT
End Sub

Sub PolyReg_Slope_T_test (r As PolyRegrec, _
 probability As Double, testValue As Double, _
 termNum As Integer, calcT As Double, tableT As Double, _
 passTest As Integer, B() As Double, StdErSlope() As Double)
 Dim p As Double, df As Double

 If probability > 1 Then
  p = .5 - probability / 200
 Else
  p = .5 - probability / 2
 End If
 df = r.sum - (r.NumTerms + 1)
 tableT = TInv(p, df)
 calcT = (B(termNum) - testValue) / StdErSlope(termNum)
 passTest = Abs(calcT) <= tableT
End Sub

Sub PolyRegCoefCI (r As PolyRegrec, probability As Double, _
 slopeHi() As Double, slopeLow() As Double, B() As Double, _
 StdErSlope() As Double)

 Dim diff As Double, p As Double
 Dim df As Double, tableT As Double
 Dim j As Integer
 Dim n As Integer

 n = r.NumTerms + 1
 If probability > 1 Then
  p = .5 - probability / 200
 Else
  p = .5 - probability / 2
 End If
 df = r.sum - (r.NumTerms + 1)
 tableT = TInv(p, df)
 For j = 1 To n - 1
  diff = tableT * StdErSlope(j)
  slopeHi(j) = B(j) + diff
  slopeLow(j) = B(j) - diff
 Next j
End Sub
```

Listing 11.4 *(Continued)*

```
Sub PolyYHatCI (r As PolyRegrec, probability As Double, _
X As Double, yHatLow As Double, yHat As Double, _
yHatHigh As Double, B() As Double, S() As Double, _
Mean() As Double)

  Dim diff As Double, p As Double
  Dim df As Double, tableT As Double
  Dim traceMat As Double, pivot As Double
  Dim tempo As Double
  Dim i As Integer, j As Integer
  Dim k As Integer, m As Integer
  Dim XX() As Double
  Dim prodMat() As Double
  Dim Xpow() As Double
  Dim n As Integer

  n = r.NumTerms + 1
  ReDim XX(n, n)
  ReDim prodMat(n, n)
  ReDim Xpow(n)

  If probability > 1 Then
   p = .5 - probability / 200
  Else
   p = .5 - probability / 2
  End If
  df = r.sum - (r.NumTerms + 1)
  tableT = TInv(p, df)

  If Not r.Inverted_S Then ' Invert matrix S
   r.Inverted_S = True
   For j = 1 To n - 1
    pivot = S(j, j)
    S(j, j) = 1
    For k = 1 To n - 1
     S(j, k) = S(j, k) / pivot
    Next k
    For k = 1 To n - 1
     If k <> j Then
      tempo = S(k, j)
      S(k, j) = 0
      For m = 1 To n - 1
       S(k, m) = S(k, m) - S(j, m) * tempo
      Next m
     End If
    Next k
   Next j
  End If

  ' Calculate yHat
  yHat = r.Intercept
  X = MyPolyFx(X)
  For i = 1 To n - 1
   Xpow(i) = X ^ i
   yHat = yHat + B(i) * Xpow(i)
  Next i

  ' Form standarized vector
  For i = 1 To n - 1
   Xpow(i) = Xpow(i) - Mean(i)
  Next i
```

Listing 11.4 *(Continued)*

```
' Form X X' matrix
For k = 1 To n - 1
 For j = 1 To n - 1
  XX(j, k) = 0
 Next j
Next k

For k = 1 To n - 1
 For j = 1 To n - 1
  XX(j, k) = XX(j, k) + Xpow(j) * Xpow(k)
 Next j
Next k

' Multiply S_Inverse and XX'
For i = 1 To n - 1
 For j = 1 To n - 1
  prodMat(i, j) = 0
  For k = 1 To n - 1
   prodMat(i, j) = prodMat(i, j) + S(i, k) * XX(k, j)
  Next k
 Next j
Next i

' Calculate trace of prodMat
traceMat = 1#
For i = 1 To n - 1
 traceMat = traceMat * prodMat(i, i)
Next i
diff = tableT * Sqr(r.ANOVA.S2 * (1 + 1 / r.sum + traceMat))
yHatLow = yHat - diff
yHatHigh = yHat + diff

yHat = MyPolyInvFy(yHat)
yHatLow = MyPolyInvFy(yHatLow)
yHatHigh = MyPolyInvFy(yHatHigh)
End Sub
```

The module file declares the following subroutines:

1. The subroutine InitializePolyReg initializes a PolyRegrec-type variable to prepare it for a new set of calculations. The parameter r is the initialized PolyRegrec-type variable. The parameter NumTerms specifies the order of the polynomial. The parameter HasMissingData specifies the missing-code flag. The parameter MissingCode indicates the value of the missing code. If the argument to parameter HasMissingData is zero, the argument for parameter MissingCode can be any value. The parameters A( ) and B( ) pass the summations matix and summations vector to be initialized by the subroutine. You must call subroutine InitializePolyReg at least once before you perform polynomial regression analysis.

2. The subroutine CalcSumPolyReg updates the statistical summations. The parameter DataMat is a matrix that supplies the observed data. The parameter r is the PolyRegrec-type variable that stores the statistical summations. The parameter NData specifies the number of observations to process. The parameters A( ), B( ), and Index( ) pass the summations matrix, summations vector, and variable-selection index array, respectively. You can call subroutine CalcSumPolyReg more than once.

3. The subroutine CalcPolyReg calculates the regression slopes, intercept, ANOVA, and other related data. The parameter r is the PolyRegrec-type variable that holds the regression matrices and vectors. The parameters A( ), B( ), S( ), Mean( ), StdErSlope( ), and TCalc( ), pass the summations matrix, summations vector, inverse summations natrix, means array, standard slope errors array, and calculated Student-t values array, respectively. You must call subroutine CalcPolyReg before calling the remaining subroutines listed here.

4. The subroutine PolyRegYHatCI calculates the projected value of the dependent variable and its confidence interval for a given value of the independent variable. The parameter r is the PolyRegrec-type variable that contains the results of polynomial regression. The parameter Probability specifies the confidence level (expressed as a percentage or a decimal). The parameter X specifies the value of the independent variable. The parameters yHat, yHatLow, and yHatHigh pass the value for the projected value and its confidence interval. The parameters B( ), S( ), and Mean( ) pass the summations vector, inverse summations matrix, and means array, respectively.

5. The subroutine PolyRegCoefCI calculates the confidence intervals for the regression slopes. The parameter r is the PolyRegrec-type variable that contains the results of polynomial regression. The parameter Probability specifies the confidence level (expressed as a percentage or a decimal). The parameter slopeHi( ) specifies the array of upper limits for the regression slope: the element slopeHi(1) corresponds to the slope for variable X, the element slopeHi(2) corresponds to the slope for X^2, and so on. The parameter slopeLow( ) specifies the array of lower limits for the regression slope: the element slopeLow(1) corresponds to the slope for variable X, the element slopeLow(2) corresponds to the slope for X^2, and so on. The parameters B( ) and StdErSlope( ) pass the summations vector and the array of standard slope errors, respectively.

6. The subroutine PolyReg_Slope_T_test tests the value of a specific regression slope. The parameter r is the PolyRegrec-type variable that contains the results of polynomial regression. The parameter Probability specifies the confidence level (expression as a percentage or a decimal). The parameter testValue specifies the tested value of the targeted regression slope. The parameter termNum is the index of the tested regression slope. The parameter calcT reports back the calculated Student-t value. The parameter tableT reports back the tabulated Student-t value. The parameter passTest reports back whether the outcome of the test is positive or negative.

7. The subroutine PolyReg_R2_T_test tests whether the value of the correlation coefficient is not zero. The parameter r is the PolyRegrec-type variable that contains the results of polynomial regression. The parameter Probability specifies the confidence level (expression as a percent or a decimal). The parameter calcT reports back the calculated Student-t value. The parameter tableT reports back the tabulated Student-t value. The parameter passTest reports back whether the outcome of the test is positive or negative.

Listing 11.5 shows the source code for the MYPOLYRG.BAS module file. This file contains the Visual Basic functions MyPolyFx, MyPolyFy, and MyPolyInvFy. The first two functions transform the values of X and Y, respectively. The function MyPolyInvFy performs the inverse transformation on the values of Y.

Listing 11.5 The source code for the MYPOLYRG.BAS module file.

```
Function MyPolyFx (X As Double)
 MyPolyFx = X
End Function

Function MyPolyFy (Y As Double)
 MyPolyFy = Y
End Function

Function MyPolyInvFy (Y As Double)
 MyPolyInvFy = Y
End Function
```

The Visual Basic test program

Let's look at the test program for the multiple regression library. Listing 11.6 shows the source code for the form associated with the TSPOLYRG.MAK program project. To compile the test program, you need to include the files POLYREG.BAS, MYPOLYREG.BAS, and STATLIB.BAS in your project file.

Listing 11.6 The source code for the form associated with the TSPOLYRG.MAK program project.

```
Dim A(10, 10) As Double
Dim S(10, 10) As Double
Dim B(10) As Double
Dim Mean(10) As Double
Dim StdErSlope(10) As Double
Dim TCalc(10) As Double
Dim Index(10) As Integer
Dim mat(15, 10) As Double

Dim r As PolyRegrec

Dim numTerms As Integer

Sub ExitMnu_Click ()
 End
End Sub

Sub Form_Load ()
 SlopeCIMnu.Enabled = False
 TestSlopesMnu.Enabled = False
 TestR2Mnu.Enabled = False
 YhatMnu.Enabled = False
End Sub

Sub PolyRegMnu_Click ()
 Dim i As Integer
 Dim numData As Integer

 mat(0, 0) = 1#
 mat(1, 0) = 1.18
 mat(2, 0) = 1.48
```

Listing 11.6 *(Continued)*

```
mat(3, 0) = 1.54
mat(4, 0) = 2#
mat(5, 0) = 2.25
mat(6, 0) = 2.6
mat(7, 0) = 2.95
mat(8, 0) = 3.25
mat(9, 0) = 3.54
mat(10, 0) = 4#

mat(0, 1) = 1#
mat(1, 1) = 1.1
mat(2, 1) = 1.2
mat(3, 1) = 1.3
mat(4, 1) = 1.4
mat(5, 1) = 1.5
mat(6, 1) = 1.6
mat(7, 1) = 1.7
mat(8, 1) = 1.8
mat(9, 1) = 1.9
mat(10, 1) = 2#

numData = 11
numTerms = 2

Index(0) = 0
Index(1) = 1

InitializePolyReg r, numTerms, False, 0, A(), B()
CalcSumPolyReg mat(), r, numData, A(), B(), Index()
CalcPolyReg r, A(), B(), S(), Mean(), StdErSlope(), TCalc()
Cls
Print "************* Polynomial Regression **********"
Print
Print "Number of points: "; Format$(r.sum, "###")
Print "R^2 = "; Format$(r.R2, "0.00000")
Print "Adjusted R^2 = "; Format$(r.R2adj, "0.00000")
Print "Intercept = "; Format$(r.Intercept, "#.####E+00")
For i = 1 To numTerms
 Print "Slope for X";
 Print Format$(i, "#");
 Print " = ";
 Print Format$(B(i), "#.####E+00")
Next i
Print
Print "************* ANOVA **********"
Print "Regression SS = "; Format$(r.ANOVA.Reg_SS, "#.####E+00")
Print "Regression df = "; Format$(r.ANOVA.Reg_df, "#.####E+00")
Print "Residual SS = "; Format$(r.ANOVA.Residual_SS, "#.####E+00")
Print "Residual df = "; Format$(r.ANOVA.Residual_df, "#.####E+00")
Print "Total SS = "; Format$(r.ANOVA.Total_SS, "#.####E+00")
Print "Total df = "; Format$(r.ANOVA.Total_df, "#.####E+00")
Print "S^2 = "; Format$(r.ANOVA.S2, "#.####E+00")
Print "F = "; Format$(r.ANOVA.F, "#.####E+00")

 ' enable remaining menu selections
 SlopeCIMnu.Enabled = True
 TestSlopesMnu.Enabled = True
 TestR2Mnu.Enabled = True
 YhatMnu.Enabled = True
End Sub

Sub SlopeCIMnu_Click ()
 Static slopeLow(10) As Double
```

Listing 11.6 *(Continued)*

```
  Static slopeHi(10) As Double
  Dim probability As Double


  probability = 95
  PolyRegCoefCI r, probability, slopeHi(), slopeLow(), B(), StdErSlope()
  Cls
  Print "************* Slope Confidence Intervals *************"
  Print
  Print "At "; Format$(probability, "##.0"); "% probability"
  For i = 1 To numTerms
   Print "Range for slope of X";
   Print Format$(i, "#");
   Print " is "; Format$(slopeLow(i), "0.00000E+00");
   Print " to "; Format$(slopeHi(i), "0.00000E+00")
  Next i
End Sub


Sub TestR2Mnu_Click ()
 Dim probability As Double
 Dim testValue As Double
 Dim passTest As Integer
 Dim calcT As Double
 Dim tableT As Double


 probability = 95
 Cls
 Print "************ Testing R^2 *************"
 Print
 Print "At "; Format$(probability, "##.0"); "% probability"
 PolyReg_R2_T_Test r, probability, calcT, tableT, passTest
 Print "H0: ";
 Print Format$(r.R2, "0.000000");
 Print " is not 0, ";
 If passTest Then
   Print "cannot be rejected"
 Else
   Print "cannot be accepted"
 End If
End Sub


Sub TestSlopesMnu_Click ()
 Dim i As Integer
 Dim probability As Double
 Dim testValue As Double
 Dim passTest As Integer
 Dim calcT As Double
 Dim tableT As Double


 probability = 95
 Cls
 Print "************ Testing Slopes *************"
 Print
 Print "At "; Format$(probability, "##.0"); "% probability"
 For i = 1 To numTerms
  Select Case i
   Case 1
   testValue = 0

   Case 2
   testValue = 1
  End Select
```

Listing 11.6 (*Continued*)

```
  PolyReg_Slope_T_test r, probability, testValue, i, calcT, tableT, passTest, B(),
StdErSlope()
  Print "H0: "; Format$(B(i), "0.00000E+00");
  Print " ?=? "; testValue;
  If passTest Then
    Print " cannot be rejected"
  Else
    Print " cannot be accepted"
  End If
 Next i
End Sub

Sub YhatMnu_Click ()
 Dim X As Double
 Dim i As Integer
 Dim probability As Double
 Dim Yhat As Double
 Dim YHatLow As Double
 Dim YHatHigh As Double

 X = mat(1, 0)
 probability = 95
 PolyYHatCI r, probability, X, YHatLow, Yhat, YHatHigh, B(), S(), Mean()
 Cls
 Print "*********** Projected Y *************"
 Print
 Print "At "; Format$(probability, "##.0"); "% probability"
 Print "X = "; Format$(X, "#####.##")
 Print "Y^ = "; Format$(Yhat, "0.00000E+00")
 Print "Range for Y^ is ";
 Print Format$(YHatLow, "0.00000E+00");
 Print " to ";
 Print Format$(YHatHigh, "0.00000E+00")
End Sub
```

The project uses a form that has a simple menu system but no controls. Table 11.2 shows the menu structure and the names of the menu items. The form has the caption "Polynomial Regression." The menu option Test has five selections to test the various regression-related calculations. Each one of these menu selections clears the form and then displays the result of some regression calculations. Thus, you can zoom in on any method by invoking its related menu selection.

TABLE 11.2 The Menu System for the TSPOLYRG.MAK Project

Menu caption	Name
&Test	TesMnu
Polynomial Regression	PolyRegMnu
Slope Confidence Intervals	SlopeCIMnu
Test Slopes	TestSlopesMnu
Test R^2	TestR2Mnu
Projected Y	YhatMnu
–	N1
&Exit	ExitMnu

The program tests the various regression-related functions using internal data. The test program declares a collection of arrays and matrices, a MLRegrec-type variable, and a simple variable in the form level. The Polynomial Regression menu selection performs the following tasks:

1. Assign the test data to the elements of matrix Mat.

2. Assign the values to the variables that store the number of data and number of terms (that is, the number of independent variables).

3. Assign the indices for the regression variables to the elements of array Index.

4. Initialize the PolyRegrec-type variable r by calling the subroutine Initialize PolyReg. The arguments for the subroutine call are the variable r, the variable numTerms, the array Index, the value FALSE, the value 0, the matrix A, and the array B. The latter two arguments indicate that the data matrix will have no missing observations.

5. Update the statistical summations by calling subroutine CalcSumPolyReg. The arguments for this function call are the matrix Mat, the variable r, the variable num-Data, the matrix A, the array B, and the array Index.

6. Calculate the regression slopes and other results by calling the subroutine CalcPolyReg. The arguments for this subroutine call are the variable r, the matrix A, the array B, the array Mean, the array StdErSlope, and the array TCalc.

7. Display the regression results, which include the coefficient of correlation, the intercept, the regression slopes, and the ANOVA table components.

The Slope Confidence Intervals menu selection calculates and displays the confidence intervals, at 95% confidence, for the regression slopes. This task involves calling subroutine PolyRegCoefCI. The program uses the arrays slopeHi and slopeLow to obtain the sought confidence intervals.

The Test Slopes menu selection tests the values for the two regression slopes at 95% confidence. The program calls subroutine PolyReg_Slope_T_test to determine if the first and second slopes are not that significantly different from the values of zero and one, respectively. The program displays the outcome of the test.

The Test R^2 menu selection tests whether or not the value of the correlation coefficient is zero (at 95% confidence). The program calls subroutine PolyReg_R2_T_test and displays the outcome of the test.

The Projected Y menu selection calculates the projected value of the dependent variable and its confidence interval (at 95% confidence) for the first row of data. The program calls subroutine PolyYHatCI and passes the arguments r, probability, Xarr( ), yHatLow, yHat, yHatHigh, B( ), S( ), and Mean( ). The program then displays the value of the dependent variables, the projected value, and the confidence interval for the projection.

Figure 11.3 shows the output of the test program for each menu selection.

```
************* Polynomial Regression **********

Number of points: 11
R^2 = 0.99650
Adjusted R^2 = 0.99518
Intercept = -2.3662E-01
Slope for X1 = 3.0566E-01
Slope for X2 = 9.0326E-01

************* ANOVA **********
Regression SS = 1.0103E+01
Regression df = 2.E+00
Residual SS = 3.5407E-02
Residual df = 8.E+00
Total SS = 1.0108E+01
Total df = 3.E+00
S^2 = 4.4259E-03
F = 1.1379E+03


************* Slope Confidence Intervals *************

At 95.0% probability
Range for slope of X1 is -1.26955E+00 to 1.88088E+00
Range for slope of X2 is 3.80453E-01 to 1.42607E+00


************ Testing Slopes *************

At 95.0% probability
H0: 3.05664E-01 ?=? 0 cannot be rejected
H0: 9.03263E-01 ?=? 1 cannot be rejected


************ Testing R^2 *************

At 95.0% probability
H0: 0.996497 is not 0, cannot be accepted


************ Projected Y *************

At 95.0% probability
X = 1.18
Y^ = 1.38177E+00
Range for Y^ is 1.22181E+00 to 1.54172E+00
```

Figure 11.3 The output of the sample Visual Basic program for polynomial regression.

The Functions Library

This chapter presents libraries of statistical and mathematical functions. These functions include the following:

- The normal probability distribution function and its inverse function
- The Student-t probability distribution function and its inverse function
- The Chi-square probability distribution function and its inverse function
- The F probability distribution function and its inverse function
- The factorial function
- The combination function
- The permutation function
- The gamma function
- The beta function
- The error function
- The error integral function
- The sine integral function
- The cosine integral function
- The Laguerre function
- The Hermite function
- The Chebyshev function
- The Bessel functions of the first and second kind

The STATLIB Library

The first library of functions in this chapter, STATLIB, has eight special functions that calculate the main four probability distribution functions and their inverses. Listing 12.1 shows the source code for the file STATLIB.BAS. This file contains the values of the constants used in the various statistical functions presented in the following sections of this chapter.

☞ Throughout the book, the underscore character is used to split wrapping lines of Visual Basic declarations and statements.

Listing 12.1 The source code for the header file STATLIB.BAS.

```
Function Chi (x As Double, df As Double) As Double
' Function will return the probability of obtaining a
' chi-squared statistic, x, at df degrees of freedom.
 Dim k As Double, xq As Double

 k = 2# / 9# / df
 xq = ((x / df) ^ (1 / 3) - (1# - k)) / Sqr(k)
 Chi = 1# - Q(xq)
End Function

Function ChiInv (x As Double, df As Double) As Double
' double will return the value of chi-squared statistic for a
' given probability, x, and df degrees of freedom.
 Dim k As Double, tempo As Double, xq As Double

 ' Check limits of x
 If x <= 0# Then x = .0001
 If x >= 1# Then x = .9999

 k = 2# / 9#
 xq = QInv(x)
 tempo = 1# - k / df + xq * Sqr(k / df)
 tempo = df * tempo ^ 3
 ChiInv = tempo ' Return sought value
End Function

Function F (x As Double, df1 As Double, df2 As Double) As Double
' Function will return the probability of obtaining an
' F statistic, x, at df degrees of freedom.
 Dim f1 As Double, f2 As Double, k As Double

 k = 2 / 9
 f1 = x ^ (1 / 3) * (1 - k / df2) - (1 - k / df1)
 f2 = Sqr(k / df1 + x ^ (2 / 3) * k / df2)
 F = 1# - Q(f1 / f2)
End Function

Function FInv (x As Double, df1 As Double, df2 As Double) _
 As Double
' Function will return the value of the F statistic for a
' given probability, x, and df degrees of freedom.
 Dim alpha As Double, beta As Double
 Dim gamma As Double, delta As Double
 Dim A As Double, b As Double, c As Double
 Dim k As Double, result As Double, xq As Double

 ' Check limits of x
 If x <= 0# Then x = .0001
 If x >= 1# Then x = .9999
```

Listing 12.1 (*Continued*)

```
 k = 2# / 9#
 xq = QInv(x)
 alpha = (1 - k / df2)
 beta = (1 - k / df1)
 gamma = 1# - (1 - k / df1)
 delta = 1# - (1 - k / df2)
 A = alpha ^ 2 - xq ^ 2 * delta
 b = -2# * alpha * beta
 c = beta ^ 2 - xq ^ 2 * gamma

 result = (-1# * b + Sqr(b * b - 4# * A * c)) / (2# * A)
 FInv = result ^ 3
End Function


Function Q (x As Double) As Double
' Function that returns the "Normal"
'    probability distribution integral
 Const twoPi = 6.283185308

 Dim result As Double
 Dim sum As Double
 Dim xp As Double
 Dim tempo As Double
 Static b(5) As Double
 Dim i As Integer

 b(1) = .31938153
 b(2) = -.356563782
 b(3) = 1.781477937
 b(4) = -1.821255978
 b(5) = 1.330274429


 tempo = 1# / (1# + .2316419 * Abs(x))
 ' Initialize summation
 sum = 0#
 xp = 1#
 ' Loop to obtain summation term
 For i = 0 To 4
  xp = xp * tempo ' Update power factor
  sum = sum + b(i) * xp
 Next i
 ' Calculate result
 result = (Exp(-x * x / 2#) / Sqr(twoPi) * sum)
 If x >= 0# Then
  Q = 1# - result
 Else
  Q = result
 End If
End Function


Function QInv (x As Double) As Double
' calculate the inverse normal.
 Dim result As Double
 Dim sum1 As Double
 Dim sum2 As Double
 Dim tempo As Double
 Dim xp As Double
 Dim i As Integer
 ' First and second coefficient array
 Static c(4)
 Static d(4)
```

Listing 12.1 (*Continued*)

```
c(1) = 2.515517
c(2) = .802853
c(3) = .010328
c(4) = 0#

d(1) = 1#
d(2) = 1.432788
d(3) = .189269
d(4) = .001308

' Check limits of x
If x <= 0# Then x = .0001
If x >= 1# Then x = .9999

If x <= .5 Then
  tempo = Sqr(Log(1 / x ^ 2))
Else
  tempo = Sqr(Log(1# / (1# - x) ^ 2))
End If
' Initialize summations
sum1 = 0#
sum2 = 0#
xp = 1#
' Start loop to calculate summations
For i = 0 To 3
  sum1 = sum1 + c(i) * xp
  sum2 = sum2 + d(i) * xp
  xp = xp * tempo ' Update power factor
Next i
' Calculate the result
result = tempo - sum1 / sum2
If x > .5 Then
  QInv = -result
Else
  QInv = result
End If
End Function

Function T (x As Double, df As Double) As Double
' Function will return the probability of
' obtaining a student-t statistic, x, at
' df degrees of freedom.
Dim xt As Double

xt = x * (1 - .25 / df) / Sqr(1 + x ^ 2 / 2 / df)
T = 1# - Q(xt)
End Function

Function TInv (x As Double, df As Double)
' Function will return the value of student-t statistic for a
' given probability, x, and df degrees of freedom.
Dim sum As Double
Dim xp As Double
Dim xq As Double
Static Pwr(10) As Double
Static term(10) As Double
Dim i As Integer

' Check limits of x
If x <= 0# Then x = .0001
If x >= 1# Then x = .9999
xq = QInv(x)
Pwr(1) = xq
```

Listing 12.1 (*Continued*)

```
' Loop to obtain the array of powers
For i = 2 To 9
 Pwr(i) = Pwr(i - 1) * xq
Next i
' Calculate the four terms
term(1) = .25 * (Pwr(3) + Pwr(1))
term(2) = (5 * Pwr(5) + 16 * Pwr(3) + 3 * Pwr(1)) / 96
term(3) = (3 * Pwr(7) + 19 * Pwr(5) + 17 * Pwr(3) - _
      15 * Pwr(1)) / 384
term(4) = (79 * Pwr(9) + 776 * Pwr(7) + 1482 * Pwr(5) - _
      1920 * Pwr(3) - 945 * Pwr(1)) / 92160#
' Initialize summation and power factor
sum = xq
xp = 1
' Loop to add terms
For i = 1 To 4
 xp = xp * df ' Update df power factor
 sum = sum + term(i) / xp
Next i
TInv = sum
End Function
```

The normal distribution

The normal distribution function implemented in the statistical library calculates the normal integral between minus infinity and a specified value x. The equations used for the calculations are as follows:

$$Q(x) = R \text{ if } R < 0$$
$$= 1 - R \text{ if } R \geq 0 \tag{12.1}$$

$$R = f(x)(b_1 t + b_2 t^2 + b_3 t^3 + b_4 t^4 + b_5 t^5) \tag{12.2}$$

$$t = \frac{1}{(1 + b_0 |x|)} \tag{12.3}$$

$$f(x) = \frac{e\left(\frac{-x^2}{2}\right)}{\sqrt{(2\pi)}} \tag{12.4}$$

where b_0 through b_5 are constants whose values appear in Listing 12.1.

The inverse normal distribution

The inverse normal distribution function implemented in the statistical library uses the following equations:

$$Q'(x) = \frac{t - P_1(t)}{P_2(t)} \tag{12.5}$$

$$P_1(t) = c_0 + c_1 t + c_2 t^2 \tag{12.6}$$

$$P_2(t) = 1 + d_1 t + d_2 t^2 + d_3 t^3 \tag{12.7}$$

$$t = \sqrt{\left(\ln\left(\frac{1}{Q^2}\right)\right)} \text{ for } 0 < Q \le 0.5$$

$$t = \sqrt{\left(\ln\left(\frac{1}{(1-Q^2)}\right)\right)} \text{ for } 0.5 < Q < 1$$

(12.8)

where c_0 through c_2 and d_1 through d_3 are constants that appear in Listing 12.1.

The Student-t distribution

The Student-t distribution function implemented in the statistical library calculates the distribution integral between minus infinity and a specified value x. The equations used for the calculations are as follows:

$$T(x,n) = 1 - Q(t)$$

$$t = \frac{x\left(1 - \frac{n}{4}\right)}{\sqrt{\frac{(1 + x^2)}{(2n)}}}$$

(12.9)

where n is the degrees of freedom.

The inverse Student-t distribution

The inverse Student-t distribution function implemented in the statistical library uses the following equations:

$$T'(x,n) = q + \frac{g_1(p)}{n} + \frac{g_2(q)}{n^2} + \frac{g_3(q)}{n^3} + \frac{g_4(q)}{n^4}$$

(12.11)

$$q = Q'(x)$$

(12.12)

where $g_1(q)$ through $g_4(q)$ are polynomials that contribute in approximating the value of the inverse Student-t distribution. Listing 12.1 contains the definition of these polynomials. The variable n represents the degrees of freedom.

The F distribution

The statistical library approximates the F distribution using the following equations:

$$F(x, n_1, n_2) = 1 - Q\left(\frac{f_1}{f_2}\right)$$

(12.13)

$$f1 = \left(\frac{1-k}{n_2}\right)x^{1/3} - \left(\frac{1-k}{n_1}\right)$$

(12.14)

$$f2 = \sqrt{\left(\frac{k}{n_1} + \frac{k}{n_2 \, x^{2/3}} \right)} \tag{12.15}$$

where $k = \dfrac{2}{9}$, and n1 and n2 are the degrees of freedom.

The inverse F distribution

The statistical library approximates the inverse F distribution using the following equations:

$$F'(x, n_1, n_2) = r^3 \tag{12.16}$$

$$r = \left(-B + \sqrt{\frac{(B^2 - 4AC)}{(2A)}} \right) \tag{12.17}$$

$$A = \alpha^2 - \Delta q^2 \tag{12.18}$$

$$B = -2\alpha\beta \tag{12.19}$$

$$C = \beta^2 - \tau q^2 \tag{12.20}$$

$$\alpha = \frac{1 - k}{n_2} \tag{12.21}$$

$$\beta = \frac{1 - k}{n_1} \tag{12.22}$$

$$\tau = 1 - \beta \tag{12.23}$$

$$\Delta = 1 - \alpha \tag{12.24}$$

$$q = Q'(x) \tag{12.25}$$

where $k = \dfrac{2}{9}$, and n_1 and n_2 are degrees of freedom.

The Mathematical Functions Library

This section presents the mathematical functions library that contains popular math functions. Listing 12.2 shows the source code for the MATHLIB.BAS file. The library supports the factorial, combination, permutation, gamma, beta, error, sine integral, cosine integral, Laguerre, Hermite, Chebyshev, and Bessel functions. The implementation of most of these functions uses an iterative or approximative approach. The implementations of the factorial, combination, permutation, Laguerre, Hermite, and Chebyshev functions yield exact values for these functions. By contrast, the implementations of the remaining functions are approximate, based on working with infinite-series polynomials that approximate the functions.

Listing 12.2 The source code for the MATHLIB.BAS file.

```
Function BesselJ (z As Double, v As Integer) As Double

 Const tolerance = .00000001
 Dim sum As Double
 Dim term As Double
 Dim i As Integer

 sum = 0
 i = 0

 Do
   term = (-z * z / 4) ^ i / factorial(i) / gamma(v + i + 1)
   i = i + 1
 Loop While Abs(term) > tolerance

 BeselJ = (z / 2) ^ v * sum
End Function

Function BesselY (z As Double, v As Integer) As Double

 BesselY = (BesselJ(z, v) * Cos(v * pi) - _
     BesselJ(z, -v)) / Sin(v * pi)
End Function

Function beta (z As Double, w As Double) As Double
 beta = gamma(z) * gamma(w) / gamma(z + w)
End Function

Function Chebyshev (x As Double, n As Integer) As Double

 Dim T0 As Double
 Dim T1 As Double
 Dim T2 As Double
 Dim i As Integer

 T0 = 1
 T1 = x
 i = 2

 If n < 0 Then
   Chebyshev = -1
 ElseIf n = 0 Then
   Chebyshev = T0
 ElseIf n = 1 Then
   Chebyshev = T1
 Else
  Do While i < n
   T2 = 2 * x * T1 - 2 * T0
   T0 = T1
   T1 = T2
   i = i + 1
  Loop
  Chebyshev = T2
 End If
End Function

Function Ci (x As Double) As Double

 Const tolerance = .00000001
 Dim sum As Double
 Dim term As Double
 Dim pow As Double
 Dim chs As Double
 Dim n As Integer
```

Listing 12.2 (*Continued*)

```
  sum = 0
  pow = x * x
  chs = -1
  n = 1

  Do
   term = chs * pow / (2 * n) / factorial(2 * n)
   sum = sum + term
   ' update components of the next term
   n = n + 1
   chs = -chs
   pow = pow * x * x

  Loop While Abs(term) > tolerance

  Ci = .5772156649 + Log(x) + sum
End Function

Function combination (m As Integer, n As Integer) As Double

  Dim factM As Double
  Dim factMN As Double
  Dim factN As Double
  Dim i As Integer

  factM = 1
  factMN = 1
  factN = 1

  If (m <= n) Or (m < 0) Or (n < 0) Then
   combination = -1 ' error code
   Exit Function
  End If

  ' calculate (m - n)!
  For i = 1 To m - n
   factMN = factMN * i
  Next i

  factM = factMN
  ' calculate m!
  For i = m - n + 1 To m
   factM = factM * i
  Next i

  ' calculate n!
  For i = 1 To n
    factN = factN * i
  Next i

  combination = factM / factMN / factN
End Function

Function Ei (x As Double) As Double

  Const tolerance = .00000001
  Dim sum As Double
  Dim term As Double
  Dim pow As Double
  Dim fact As Double
  Dim n As Integer

  sum = 0
  pow = x
```

Listing 12.2 (*Continued*)

```
fact = 1
n = 1

Do
 term = pow / n / fact
 sum = sum + term
 ' update components of the next term
 n = n + 1
 fact = fact * n
 pow = pow * x

Loop While Abs(term) > tolerance

Ei = .5772156649 + Log(x) + sum
End Function

Function erf (x As Double) As Double
' error function
Dim t As Double
Dim sum As Double
Dim i As Integer
Static a(5)

t = 1 / (1 + .32759 * x)
a(0) = 0
a(1) = .254829592
a(2) = -.284496736
a(3) = 1.424143741
a(4) = -1.453152027
a(5) = 1.061405429

sum = 0
For i = 5 To 0 Step -1
 sum = (sum + a(i)) * t
Next i

erf = 1 - sum / Exp(x * x)
End Function

Function factorial (n As Integer) As Double
 Dim result As Double
 Dim i As Integer
 result = 1

 If n > -1 Then
  For i = 1 To n
   result = result * i
  Next i
 End If
 factorial = result
End Function

Function gamma (x As Double) As Double

 Dim sum As Double
 Dim i As Integer
 Static c(26) As Double
 c(0) = 0
 c(1) = 1
 c(2) = .5772156649
 c(3) = -.6558780715
 c(4) = -.042002635034
 c(5) = .16665386113
```

Listing 12.2 *(Continued)*

```
c(6) = -.0421977345555
c(7) = -.0096219715278
c(8) = .00721894324666
c(9) = -1.1651675918591E-03
c(10) = -2.152416741149E-04
c(11) = 1.280502823882E-04
c(12) = -2.01348547807E-05
c(13) = -1.2504934821E-06
c(14) = .000001133027232
c(15) = -2.056338417E-07
c(16) = .000000006116095
c(17) = 5.0020075E-09
c(18) = -1.1812746E-09
c(19) = 1.043427E-10
c(20) = 7.7823E-12
c(21) = -3.6968E-12
c(22) = .00000000000051
c(23) = -2.06E-14
c(24) = -5.4E-15
c(25) = 1.4E-16
c(26) = 1E-16

sum = 0

For i = 26 To 0 Step -1
 sum = (sum + c(i)) * x
Next i

gamma = 1 / sum
End Function

Function Hermite (x As Double, n As Integer) As Double

 Dim H0 As Double
 Dim H1 As Double
 Dim H2 As Double
 Dim i As Integer

 H0 = 1
 H1 = 2 * x
 i = 2

 If n < 0 Then
  Hermite = -1
 ElseIf n = 0 Then
  Hermite = H0
 ElseIf n = 1 Then
  Hermite = H1
 Else
  Do While i < n
   H2 = 2 * x * H1 - 2 * i * H0
   H0 = H1
   H1 = H2
   i = i + 1
  Loop
  Hermite = H2
 End If
End Function

Function Laguerre (x As Double, n As Integer) As Double

 Dim L0 As Double
 Dim L1 As Double
```

Listing 12.2 *(Continued)*

```
Dim L2 As Double
Dim i As Integer

L0 = 1
L1 = 1 - x
i = 2

If n < 0 Then
 Laguerre = -1
ElseIf n = 0 Then
 Laguerre = L0
ElseIf n = 1 Then
 Laguerre = L1
Else
 Do While i < n
  L2 = (1 + 2 * i - x) * L1 - i * i * L0
  L0 = L1
  L1 = L2
  i = i + 1
 Loop
 Laguerre = L2
End If
End Function

Function Legrendre (x As Double, n As Double) As Double

 Dim L0 As Double
 Dim L1 As Double
 Dim L2 As Double
 Dim i As Integer

 L0 = 1
 L1 = x
 i = 2

 If n < 0 Then
  Legrendre = -1
 ElseIf n = 0 Then
  Legrendre = L0
 ElseIf n = 1 Then
  Legrendre = L1
 Else
  Do While i < n
   L2 = ((2 * i + 1) * L1 - i * L0) / (i + 1)
   L0 = L1
   L1 = L2
   i = i + 1
  Loop
  Legrendre = L2
 End If
End Function

Function permutation (m As Integer, n As Integer) As Double

 Dim factM As Double
 Dim factMN As Double
 Dim i As Integer

 factM = 1
 factMN = 1

 If (m <= n) Or (m < 0) Or (n < 0) Then
   permutation = -1 ' error code
```

Listing 12.2 (*Continued*)

```
   Exit Function
End If

' calculate (m - n)!
For i = 1 To (m - n)
 factMN = factMN * i
Next i

factM = factMN
' calculate m!
For i = m - n + 1 To m
 factM = factM * i
Next i

permutation = factM / factMN
End Function

Function Si (x As Double) As Double

 Const tolerance = .00000001
 Dim sum As Double
 Dim term As Double
 Dim pow As Double
 Dim chs As Double
 Dim n As Integer

 sum = 0
 pow = x
 chs = 1
 n = 0

 Do
  term = chs * pow / (2 * n + 1) / factorial(2 * n + 1)
  sum = sum + term
  ' update components of the next term
  n = n + 1
  chs = -chs
  pow = pow * x * x
 Loop While Abs(term) > tolerance

 Si = sum
End Function
```

The combination and permutation functions

The implementation of the factorial function uses a For loop that evaluates the factorial function defined by the following equation:

$$n! = n\,(n-1)\,(n-2) \dots 2\,1 \qquad (12.26)$$

The following equations define the combination and permutation functions, respectively:

$$_mC_n = \frac{m!}{[(m-n)!\,n!]} \qquad (12.27)$$

$$_mP_n = \frac{m!}{(m-n)!} \qquad (12.28)$$

The gamma and beta functions

The implementation of the gamma uses a series expansion given by the following equation:

$$\frac{1}{\Gamma(x)} = \Sigma c_k \, x^k \tag{12.29}$$

$$\text{for } k = 1 \text{ to } 26$$

where c_1 through c_{26} are constants shown in Listing 12.2. The beta function uses the gamma function to yield a value, as shown here:

$$B(z,w) = \frac{\Gamma(z) \, \Gamma(w)}{\Gamma(z + w)} \tag{12.30}$$

The error function

The error function has several approximations. The mathematical library MATHLIB uses the following one:

$$\text{erf}(x) = 1 - (a_1 t + a_2 t^2 + a_3 t^3 + a_4 t^4 + a_5 t^5) \, e^{-y} \tag{12.31}$$

$$t = \frac{1}{(1 + 0.32759 \, x)} \tag{12.32}$$

where $y = x^2$ and a_1 through a_5 are constants shown in Listing 12.2.

The sine and cosine integral functions

The mathematical library calculates the values for the sine and cosine integrals using the following series expansions:

$$Si(x) = \sum \frac{[(-1)^n x^{(2n + 1)}]}{[(2n + 1)(2n + 1)!]} \tag{12.33}$$

$$Ci(x) = \Gamma + \ln(x) + \sum \frac{(-1)^n x^{2n}}{[(2n)(2n)!]} \tag{12.34}$$

where Γ is the Euler constant. These summations are taken for $n = 1$ and upward (until the absolute value of the evaluated term falls below a minimum limit).

The Laguerre, Hermite, and Chebyshev polynomials

The mathematical library offers functions to evaluate the Laguerre, Hermite, and Chebyshev polynomials. The code uses loops to calculate these polynomials based on recursive formulas. The recursive equations for the Laguerre polynomial are as follows:

$$L_0(x) = 1 \tag{12.35}$$

$$L_1(x) = 1 - x \tag{12.36}$$

$$L_{n+1}(x) = (1 + 2n - x)L_n(x) - n^2 L_{n-1}(x) \tag{12.37}$$

The recursive equations for the Hermite polynomial are as follows:

$$H_0(x) = 1 \tag{12.38}$$

$$H_1(x) = 2x \tag{12.39}$$

$$H_{n+1}(x) = 2xH_n(x) + 2nH_{n-1}(x) \tag{12.40}$$

As for the Chebyshev polynomial, the mathematical library uses the following non-recursive equation to evaluate the polynomial:

$$T_n(x) = \cos(n \cos^{-1} x) \tag{12.41}$$

The Bessel functions

The mathematical library implements the Bessel functions of the first and second kinds. The following equation provides a series expansion for the Bessel function of the first kind:

$$J_n(x) = \left(\frac{x}{2}\right)^n \sum \frac{\left(\frac{-z2}{4}\right)^k}{[k!\Gamma(n + k + 1)]} \tag{12.42}$$

This summation is taken for $k = 1$ and upward (until the absolute value of the evaluated term falls below a minimum limit).

The mathematical library evaluates the Bessel function of the second kind using the following equation:

$$Y_n(x) = \frac{[J_n(x) \cos(n\pi) - J_{-n}(x)]}{\sin(n\pi)} \tag{12.43}$$

Index

Illustrations are indicated by **boldface.**

ABOUT THE AUTHOR

Namir C. Shammas is the author of more than 20 books on programming topics for McGraw-Hill, Prentice-Hall, Wiley, and M&T. His articles and columns have appeared in such journals as *Computer Language*, *Dr. Dobb's*, and *BYTE*. Shammas holds a Master of Science in chemical engineering from the University of Michigan.

DISK WARRANTY

This software is protected by both United States copyright law and international copyright treaty provision. You must treat this software just like a book, except that you may copy it into a computer in order to be used and you may make archival copies of the software for the sole purpose of backing up our software and protecting your investment from loss.

By saying "just like a book," McGraw-Hill means, for example, that this software may be used by any number of people and may be freely moved from one computer location to another, so long as there is no possibility of its being used at one location or on one computer while it also is being used at another. Just as a book cannot be read by two different people in two different places at the same time, neither can the software be used by two different people in two different places at the same time (unless, of course, McGraw-Hill's copyright is being violated).

LIMITED WARRANTY

Windcrest/McGraw-Hill takes great care to provide you with top-quality software, thoroughly checked to prevent virus infections. McGraw-Hill warrants the physical diskette(s) contained herein to be free of defects in materials and workmanship for a period of sixty days from the purchase date. If McGraw-Hill receives written notification within the warranty period of defects in materials or workmanship, and such notification is determined by McGraw-Hill to be correct, McGraw-Hill will replace the defective diskette(s). Send requests to:

McGraw-Hill, Inc.
Customer Services
P.O. Box 545
Blacklick, OH 43004-0545

The entire and exclusive liability and remedy for breach of this Limited Warranty shall be limited to replacement of defective diskette(s) and shall not include or extend to any claim for or right to cover any other damages, including but not limited to, loss of profit, data, or use of the software, or special, incidental, or consequential damages or other similar claims, even if McGraw-Hill has been specifically advised of the possibility of such damages. In no event will McGraw-Hill's liability for any damages to you or any other person ever exceed the lower of suggested list price or actual price paid for the license to use the software, regardless of any form of the claim.

McGRAW-HILL, INC. SPECIFICALLY DISCLAIMS ALL OTHER WARRANTIES, EXPRESS OR IMPLIED, INCLUDING, BUT NOT LIMITED TO, ANY IMPLIED WARRANTY OF MERCHANTABILITY OR FITNESS FOR A PARTICULAR PURPOSE.

Specifically, McGraw-Hill makes no representation or warranty that the software is fit for any particular purpose and any implied warranty of merchantability is limited to the sixty-day duration of the Limited Warranty covering the physical diskette(s) only (and not the software) and is otherwise expressly and specifically disclaimed.

This limited warranty gives you specific legal rights; you may have others which may vary from state to state. Some states do not allow the exclusion of incidental or consequential damages, or the limitation on how long an implied warranty lasts, so some of the above may not apply to you.

Disk Information

The double density 3.5" disk bound with this book contains the Visual Basic module files and test programs found in these pages, (listed again below for your convenience). Note that a working copy of the Visual Basic software is needed in order to implement the code.

ANOVA.BAS	ROOT.FRM
BASTAT.BAS	STATLIB.BAS
BLANK.MAK	TSANOVA.MAK
DERIV.BAS	TSANOVA.FRM
INTEGRAL.BAS	TSBASTAT.FRM
INTEGRAL.MAK	TSBASTAT.MAK
INTERP.BAS	TSDERIV.FRM
INTERP.FRM	TSDERIV.MAK
LINREG.BAS	TSINTEG.FRM
MATHLIB.BAS	TSINTEG.MAK
MATVECT.BAS	TSLINREG.FRM
MLREG.BAS	TSLINREG.MAK
MYDERIV.BAS	TSMAT.FRM
MYINTEG.BAS	TSMAT.MAK
MYLINREG.BAS	TSMLREG.FRM
MYMLREG.BAS	TSMLREG.MAK
MYODE.BAS	TSODE.FRM
MYOPTIM.BAS	TSODE.MAK
MYPOLYRG.BAS	TSOPTIM.FRM
MYROOT.BAS	TSOPTIM.MAK
ODE.BAS	TSPOLYRG.FRM
OPTIM.BAS	TSPOLYRG.MAK
POLYREG.BAS	TSROOT.MAK
ROOT.BAS	TSROOT.FRM

Important